INTELLIGENT SUPERVISORY CONTROL
A Qualitative Bond Graph Reasoning Approach

WORLD SCIENTIFIC SERIES IN ROBOTICS AND INTELLIGENT SYSTEMS

Editor-in-Charge: C J Harris (*University of Southampton*)
Advisor: T M Husband (*University of Salford*)

World Scientific Series in Robotics and Intelligent Systems – Vol. 14

INTELLIGENT SUPERVISORY CONTROL

A Qualitative Bond Graph Reasoning Approach

HANG WANG & DEREK LINKENS

Univ. of Sheffield, UK

World Scientific

Singapore • New Jersey • London • Hong Kong

Published by

World Scientific Publishing Co. Pte. Ltd.

P O Box 128, Farrer Road, Singapore 912805

USA office: Suite 1B, 1060 Main Street, River Edge, NJ 07661

UK office: 57 Shelton Street, Covent Garden, London WC2H 9HE

Library of Congress Cataloging-in-Publication Data
Wang, Hang.
   Intelligent supervisory control: a qualitative bond
graph reasoning approach / Hang Wang & Derek Linkens.
       p. cm. -- (World Scientific series in robotics and artificial intelligence; vol. 14)
      Includes bibliographical references and index
      ISBN 9810226586
   1. Intelligent control systems.   I. Linkens, D. A. II. Title.   III. Series
   TJ217.5.W36   1996
   629.8/9 -- dc20   96-14915
                                      CIP

British Library Cataloguing-in-Publication Data
A catalogue record for this book is available from the British Library.

Printed in Singapore.

SERIES EDITOR'S FOREWORD

The classical approach to control design is to develop plant models based on linear analytic methods and synthesize control laws that encompass plant nonlinearities, modelling error, and uncertainty through inherent robustness. Unfortunately, robustness is achieved at the expense of performance. Future system requirements demand high performance, availability, fault tolerance and cost effectiveness. Additionally, many processes are of such complexity as to defy conventional modelling and control methodologies. Current developments in knowledge-based systems, learning systems, qualitative reasoning, etc and associated soft paradigms offer real opportunities for radical innovation and significant improvements in product quality assurance and fault tolerance, as well as the means of dealing with total systems management by considering qualitative as well as quantitative aspects simultaneously in an intelligent manner. Historically controller design has focused on the low level servo-sensor aspects of systems control, yet greater system benefits occur by effective management or supervision. Whereby, system controller gains etc are tuned by higher level criteria/techniques. This book makes a significant and original contribution to the subject of intelligent control, by developing both qualitative modelling via bond graphs and an associated methodology of intelligent supervisory control.

Christopher J. Harris
University of Southampton
UK

PREFACE

Intelligent systems and control has become a focus of much interest in recent years. This has been partly due to a recognition that quantitative methods in science and engineering provide only partial insight and design methods when dealing with complex systems. This is summed up in the well-known statement by L Zadeh in 1965 "...as the complexity of a system increases, our ability to make precise and yet significant statements about its behaviour diminishes until a threshold is reached beyond which precision and significance (or relevance) become almost mutually exclusive characteristics".

A number of paradigms have been developed within the area of intelligent systems. The main ones are fuzzy logic, neural networks, expert systems, genetic algorithms and qualitative reasoning. Attempts are being made increasingly to integrate these "tools" into comprehensive frameworks or to merge two or more of the techniques in a synergetic manner. The most notable progress to date are the advances in neuro fuzzy logic and control. In contrast to this approach, this book seeks to use two less well-known paradigms in a closely integrated manner to tackle increasingly difficult design and operational tasks which are necessary for systems to be classified as autonomous in nature.

In this book, a knowledge representation and utilisation methodology integrating qualitative reasoning and bond graphs is developed to construct intelligent supervisory control systems. Qualitative reasoning is a powerful model-based reasoning method while bond graphs are a formal modelling language of dynamic systems. Their integration, qualitative reasoning on bond graphs, results in a problem-solving approach in Artificial Intelligence, in which qualitative reasoning is used as the general reasoning strategy and bond graphs are employed as the knowledge representation.

After an Introduction and Review chapter, a systematic modelling procedure based on qualitative bond graphs is presented in Chapter 3. A controller design method is then developed in Chapter 4 to derive control algorithms from qualitative bond graph models. Because of uncertainty in process dynamics, an auto-tuning scheme is proposed in Chapter 5 to adjust the controllers in order to meet performance criteria and adapt to system changes. For deeper levels of intelligent control, a fault diagnosis mechanism is built in Chapter 6 to localise system faults, and an additional measurement suggestion method is developed for the refinement of diagnosis results. To move further towards autonomy, an automatic planner is proposed to generate the operation sequences for system start-up, shut-down, and emergency measures to help human operators operate systems safely. All of these applications are combined together in Chapter 7 via a management mechanism to construct a supervisory control system.

The implementation of this integrated supervisory control system is illustrated throughout the chapters by experiments with single-variable and multi-variable coupled tanks liquid level control rigs, operating in real-time using software code written in Modula-2.

CONTENTS

CHAPTER 1

INTRODUCTION

1.1 PREAMBLE

The principal purposes of industrial control are high product quality, high flexibility, economic efficiency, and safety. Early work in control engineering concentrated on the first three objectives. The development of classical linear control theories made it possible to design controllers according to specified performance criteria such as stability and precision. The use of state-space equations as system models led to the approaches of optimal control, multivariable control, and Lyapunov stability etc. The progress of computer techniques encouraged the study of digital control methods. The program which characterises a digital controller can be modified to accommodate design changes and adaptive performances without variations in the hardware, so that a controller can be built with higher flexibility and less cost. Computer-Aided Design (CAD) of control systems further curtailed the time and reduced the cost for controller development. All these improvements in control engineering were based on the precise mathematical representations of systems.

As industrial plants grow larger and more complex, control system design becomes more and more difficult. Establishing precise mathematical models for complex systems is costly and time-consuming. Design techniques and algorithms involved in CAD facilities are very intensive so that only a few specialists can utilise the CAD toolkits. On the other hand, the complexity of plants is so great that human operators cannot handle it effectively. With the increasing complexity, there are enormous possibilities for plants to go wrong. It is very difficult to predict all the possible faults and lay down contingency plans to cope with them. Operating a plant safely has become a big challenge for control engineering. Conventional mathematical

1

models do not provide the knowledge about the interrelations of plant structures and functions, which is essential for undertaking the tasks of fault diagnosis, plant start-up, and shut-down. As a result, quantitative techniques cannot satisfy the requirement of safety. In these circumstances, techniques of artificial intelligence (AI) can be used to deal with the complexity to reduce the burdens on humans in both controller design and plant operation.

AI is a branch of computer science, which is concerned with the automation of human intelligence. AI has a number of important, helpful features which allow control engineering to tackle system complexity. First, it can derive solutions with inexact or incomplete information. In the real world, very few problems can be well defined easily. AI techniques are able to imitate human thinking to infer the solutions by integrating incomplete information and associative relationships. Although the solutions may be neither exact nor optimal, they at least offer the "sufficient" answers when detailed mathematical models are too expensive or not available. Prototypes of control systems can be constructed quickly and easily. Second, AI uses the representational formulations in which domain-specific knowledge can be improved, updated, and corrected by the user or the machine itself without variations on other parts of the control systems. The capability of control systems can be improved efficiently along with the increased knowledge obtained from operation or simulation experience. Control systems become more flexible in coping with plant and environment changes. Finally, AI uses symbolic computations to reason about heuristic knowledge, instead of solving numerical equations. Through this symbolic reasoning, what will happen in plants can be inferred, and how plants work and how things fail can be explained. Plant operators can manage plants better with the guidance suggested by computers.

As a branch of computer science, AI approaches must be based on principles applicable to computers. These principles include the programmable knowledge representation in which solutions to problems can be derived from the facts stored therein, and the inference mechanisms needed to apply the knowledge to resolve problems. This book is thus devoted to these two aspects: developing a knowledge representation suitable for dynamic physical systems, and developing the inference mechanisms required for industrial control tasks. Its goal is to build an intelligent supervisory control system by integrating AI techniques with control engineering.

1.2   MOTIVATION AND BACKGROUND

The research in this book was originally motivated by the aim to build an on-line supervisory control system to help human operators manage industrial plants. This system should be applicable generically for components involved in industrial systems, such as electrics, mechanics, and thermodynamics. Also, it should be easily utilised by users who are not familiar with control engineering or AI techniques.

The main objectives of operators are first to start up a process, monitor its behaviour, maintain it close to the required steady state, and, if necessary, shut down the process when serious faults have been detected. Usually, operators handle a process through monitoring several key variables which are significant to the process situation and easily observed. In doing so, numerical values of these variables are translated into qualitative semantic terms such as "overshoot is too big", "steady-state error is normal", and so on. When the process is regarded as abnormal, operators will determine the possible reasons and decide how to correct the abnormal behaviour through analysing the interactions between process components. Usually, operators do not operate a process by resolving mathematical equations. Instead, they integrate all the process information, either complete or incomplete, with the knowledge about the process to infer the solutions for engineering problems. If a supervisory control system can simulate this human-thinking, then, ideally, this system can operate in a similar manner to that of human operators. Such a supervisory system should be able to handle and express the qualitative information, and have knowledge about the process structure.

In order to meet these requirements, the first problem to be considered is how to acquire the structural information systematically and represent it in a form applicable to computers. The "Qualitative Reasoning Environment for Modelling and Simulation (QREMS)" [Linkens *et al*, 1991] provided an economic way to solve this problem. QREMS combined AI techniques, qualitative reasoning, and the formal modelling language of dynamic bond graphs, to build an automatic modelling and simulation system. In this approach, qualitative models of processes were constructed by studying their internal mechanisms and represented by the constitutive physical laws of process components and their interconnections. General properties of a process were then analysed through reasoning about the behaviour of individual components of the process and their functional relations, using qualitative reasoning techniques. QREMS provided not only a systematic knowledge acquisition method but also a method for operating the qualitative information. This idea, qualitative bond graph reasoning, is employed as the basis for the supervisory control system developed in this book.

After the knowledge acquisition method has been decided, the next problem to be considered is how to represent the knowledge appropriately for the various tasks in supervisory control. A model for feedback control is required to describe the input-output relations of a process. Internal physical variables of the process are not necessary for the feedback control task, so this model does not need to represent them. On the other hand, precision is one of the most important requirements for a feedback controller. Therefore, this model should be able to accept and describe the numerical data about the process so that an accurate controller can be designed according to the precise representation of the input-output relations. A model for fault diagnosis is required to describe the locations of process components and the interactions between the variables of these components so that possible faults can be localised through analysing the interrelations of the component situations and abnormal behaviour observed. This model does not need to describe the numerical details crucial for feedback control, since fault diagnosis usually relies on cause-effect inference rather than on numerical calculations. Similarly, the model for the derivation of start-up and shut-down sequences should be able to describe the structural information and, further, the behaviour of switches so that the effect of operating a switch can be analysed.

As discussed above, supervisory control is composed of various types of reasoning in control engineering related to different aspects of knowledge about a process. Therefore, an appropriate model for supervision has to be built independently, rather than aiming at specific control tasks, so that it can involve all the necessary knowledge for a supervisory control system. Further, this model should represent both qualitative and quantitative information. In this book, a bond-graph-based representation, inspired by the approach of QREMS, is proposed for the supervision model. A set of qualitative equations representing components' physical variables, parameters, and their functional relations can be stated directly from a bond graph model. Also, these equations can be abstracted to represent the relations between input and output variables directly, and allow numerical values of process parameters and variables to be inserted.

With respect to the implementation of the real-time supervisory control system, the first aspect to be considered is its architecture. A real-time system has to respond to events which cannot be determined in advance, and thus cannot be programmed into the system. Therefore, the reasoning processes of the supervisory system should not be a fixed sequence; instead, an automatic supervisor is necessary for the system to decide suitable reasoning strategies for various events. Also, this system should be easily modified to keep it efficient. The knowledge base, the inference mechanisms for different control tasks, and the automatic supervisor must be able to be validated individually without affecting each other. In other words, functions for different tasks of a supervisory control system should be programmed independently. Thus, the architecture of the supervisory system is clear. It has a knowledge base separated from the inference mechanisms, and a system supervisor at the higher level to manage the inference mechanisms of different control tasks. In this book, the inference mechanisms include a feedback controller, an auto-tuning mechanism for the controller, a fault diagnosis mechanism, a planner for operation sequence derivation, and a performance monitor.

Another important requirement for the real-time supervisory control system is that the solutions to problems cannot afford to be ambiguous. One cause of ambiguity is the information loss in the qualitative abstraction of numerical data. Ambiguity is unacceptable for the real-time supervisory system because such a system is used to help the human operators make immediate decisions rather than confuse them. In the feedback control aspect, the variables concerned are the process input and output. This book develops a method to merge the numerical measurement of process output into qualitative control algorithms to infer a certain value for the process input. In the fault diagnosis aspect, all the internal variables of a process need to be analysed. However, it is difficult and inefficient to measure or infer the numerical values for all the internal variables. Therefore, a set of refined qualitative descriptors and operations is defined to avoid possible ambiguity.

To summarise, the purpose of this book is to develop an integrated real-time supervisory control system based on qualitative bond graph reasoning: starting with the knowledge acquisition and representation methods and covering the aspects of feedback control, auto-tuning, fault diagnosis, and the operation sequence derivation.

1.3   OVERVIEW OF THE BOOK

The remainder of this book is organised in the following way:

Chapter 2 reviews previous research in intelligent supervisory control, qualitative reasoning, and bond graphs. Observations on these approaches are made to identify their difficulties and the possible solutions.

Chapter 3 first compares two qualitative approaches, fuzzy set theory and qualitative reasoning, to remove the confusions between them; and introduces the basic bond graph modelling language. A domain-independent qualitative representation of bond graphs is proposed. A general modelling procedure is developed to model dynamic systems from their physical structures. Another model simplification method is developed to abstract input-output relations from a bond graph model for the purpose of controller design. All these methodologies are employed to build the knowledge base for the supervisory control system.

Chapter 4 presents a controller design method which can derive both single-input-single-output (SISO) and multi-input-multi-output (MIMO) control algorithms directly from simplified qualitative bond graph models. These control algorithms allow the process parameters to be inserted to improve their precision. A method of merging the observed numerical system outputs into the qualitative control algorithms is developed to further improve controllers' precision and to avoid ambiguities.

Chapter 5 presents an auto-tuning scheme for the hybrid qualitative and quantitative controllers, using a pattern recognition approach. A performance monitor is built to monitor the system behaviour. Controllers can be adjusted according to the monitoring results to meet specified performance criteria and adapt to process or environment changes.

Chapter 6 illustrates the fault diagnosis method based on qualitative bond graph models. A set of qualitative descriptors and their operations are defined for fault diagnosis. General procedures for localising the faults of components, controllers, power drivers, and sensors are developed. Fault candidates are represented by their locations and fault types. A suggestion method for additional measurements is also developed to refine the diagnosis results. This method needs no extra fault models, so it can be performed easily.

Chapter 7 combines all the techniques developed in Chapters 3 to 6 to construct a supervisory control system. The architecture for this supervisory control system is described. A management mechanism is designed to choose suitable strategies for coping with various system situations. Further, this chapter extends qualitative bond graphs to model switches and represent their behaviour. Based on this extended knowledge representation, procedures for generating the sequences for system start-up, shut-down, and emergency measures are developed to help the operator make decisions.

Chapter 8 presents conclusions drawn from the research in the preceding chapters, and makes recommendations for further work in this area.

Implementations of these methods are illustrated by experiments on SISO and MIMO coupled tanks liquid level control rigs, operating in real-time using Modula-2 software code. The experiments in each subject are described and discussed respectively in each chapter.

CHAPTER 2

REVIEW OF PREVIOUS WORK

This book aims to integrate the techniques of qualitative reasoning and bond graph modelling to develop an intelligent supervisory control system. Significant approaches to supervisory control, qualitative reasoning, and bond graphs are reviewed respectively in this chapter. The reasons for the use of qualitative reasoning and bond graphs can thus be elicited through observations about these approaches.

2.1 SCOPE OF INTELLIGENT SUPERVISORY CONTROL

Intelligent supervisory control is an integration of the disciplines of control engineering and artificial intelligence (AI). Intelligent control theory utilises the powerful high-level decision-making of the digital computer with advanced mathematical modelling and synthesis techniques of system theory to produce a unified approach suitable for engineering needs [Saridis, 1983].

The basic concepts and methods of integrating control theory with AI were formed around 1970. The original purpose of this category was to transfer, as much as possible, the designer's and human operator's intelligence to a machine controller [Fu, 1971]. To achieve this purpose, it was suggested that an intelligent controller should be composed of two parts: primary and supervisory. Activities requiring lower intelligence such as data collection, routine decision, and on-line computations, were assigned to the primary controller. On the other hand, decisions requiring relatively higher intelligence such as the recognition of complex environmental situations, setting subgoals for the primary controller, and correcting improper decisions made by the primary controller, were allocated to the supervisory controller.

One of the earliest implementations in intelligent supervisory control was the Stanford Research Institute (SRI) robot vehicle [Nilsson, 1969]. The robot was propelled by two stepper motors independently driving a wheel on either side of the vehicle. The movement of the vehicle was guided by a TV camera mounted on a movable head. The specification of the system required the vehicle to rearrange (by pushing) simple objects in its environment, and the system accomplished the specified tasks by performing a sequence of elementary actions, such as wheel motions and camera readings. The controller in this system was composed of three major components: 1) models of the system and its environment, 2) problem solver, and 3) primary controller. The problem solver utilised the information stored in the models to determine what sequence of actions could cause the controlled process to be in the desired state, and the commands given by the problem solver were then executed by the primary controller.

A generalised hierarchical architecture for intelligent robotic control was proposed by Saridis in 1977 [Saridis, 1977; 1983]. The objective of this approach was to develop an intelligent system which can provide all the features of primary controllers (e.g. stability, controllability, and precision) and human operators (e.g. flexibility, learning capability, and ease of communication). For this purpose, the system's structure was composed of three levels: *organisation* level, *coordination* level, and *execution* level, ordered according to the principle of "increasing precision with decreasing intelligence" [Saridis, 1989]. The highest level is an automatic *organiser*, which accepts and interprets the input commands and related feedback from the system, defines the task to be executed, and segments it into subtasks in their appropriate order of execution. The implementation of the organiser was obtained by a syntax-directed translation schema which generates speech recognition algorithms and an organising schema which organises the required tasks. The *coordination* level receives instructions from the organiser and feedback information from the process for each subtask to be executed, and coordinates the execution at the lowest level. A fuzzy automaton was suggested to implement the coordinator. A finite state machine was designed to select one particular subtask from a library, using a learning procedure. The *execution* level consisted of several controllers designed for effective control using an approximation theory of optimal control.

A wider range of intelligent supervisory control was represented by Åström *et al* [1986]. They observed that the actual implementation of PID control often incorporates a substantial amount of heuristic logic, which is even more important in multivariable and self-tuning regulators. Further, they stated that knowledge representation is a key issue in intelligent control systems. Knowledge in their system was described as if-then rules, and the rulebase was structured in groups of knowledge sources that contain rules about the same subject. Necessary functions to perform intelligent control were grouped into knowledge sources as follows:

Main control:	minimum variance control; minimum variance supervisor; ringing detector; degree supervisor.
Basic control:	PID control.
Estimation:	parameter estimation; estimation supervisor; excitation supervisor; perturbation signal generator; jump detector.
Self-tuning:	self-tuning regulation.
Learning:	set regulator parameters; smooth-and-store regulator parameters; test scheduling hypothesis.
Main monitor:	stability supervisor; compute means and variances.

The database for the architecture stored process data to support technical audit and learning, and was structured in event lists to deal with time-varying aspects, and hypotheses lists with different levels of abstraction.

Similarly to this approach, most existing supervisory control methods have been built in terms of smart auto-tuning of PID controllers and supervision of these adaptive controllers. These methods can be classified with respect to the usage of expert system techniques. The best known example is the Foxboro EXACT [Kraus and Myron, 1984], which was focused on control heuristics but implemented via conventional techniques based on pattern identification of transients in the control error. Liu *et al* [1987] described a supervisory control structure using heuristic rules for the supervision of adaptive controllers. Porter *et al* [1987] proposed an expert control system for PI controllers based on step response analysis using expert system techniques. The parameters in a PI controller were adjusted by heuristic tuning rules according to the qualitative transient-response characteristics of a process, e.g. monotone, oscillatory, no overshoot, etc. An architecture for integrated process supervision, consisting of primary control, adaptive control, and fault diagnosis, was developed by Leitch and Quek [1992]. A boundary detection mechanism was used to monitor system behaviour according to supervisory cost functions. The cost functions were evaluated by comparing the desired behaviour, obtained through the simulation of a reference model, with the observed behaviour. Then, the system supervisor could schedule the generic control regimes to cope with various system behaviour classified by the cost functions.

Another method of representing the supervisory heuristic rules has been implemented using fuzzy logic theory. In de Silva and Macfarlane [1989], a set of linguistic tuning rules of a PID controller was adopted to adjust the controller. Perng and Chang [1993] combined a speed-up linear controller and a linear regulator, designed via conventional linear control theory, using a fuzzy supervisory controller to regulate a non-linear system. The commercial supervisory controller developed by Omron Electronics [Infelise, 1991] combines an advanced PID algorithm and fuzzy control rules. The PID controller is employed in the start-up and steady-state operation periods, while the fuzzy logic controller is activated to correct the PID control action when an external disturbance occurs.

Supervisory control was first viewed as an aspect of knowledge-based control by Årzén [1988]. His research was focused on what process knowledge is needed in order to automatically tune and supervise an arbitrary controller, and how this knowledge should be represented and implemented. A blackboard architecture [Haton, 1983] was adopted to decompose a problem into subtasks which are implemented as separate knowledge sources that can be rule-based with different inference strategies or written in terms of procedures. The blackboard was used as the reasoning model, which is available to different cooperating knowledge sources. The knowledge sources contain the heuristic logic surrounding the involved algorithms, and can be seen as specialists on different subtasks of the problem such as controller design, model validation, and different monitoring aspects, etc. As data or a fact appears on the blackboard, the knowledge sources check to see if they can use that fact to fire some of their rules. If so, a knowledge source may add data to the blackboard, add or modify hypotheses, or update the facts. The operation of the knowledge-based controller involves the activation of different knowledge sources both in sequence and in parallel. The rule-based scheduler selects and schedules knowledge sources for operations. A knowledge source runs until it has to wait for some information or it is finished. Three different strategies for combining knowledge sources into sequences have been implemented. The first is to use primitives that let knowledge sources activate or deactivate each other. The second is to have a number of pre-stored sequences, such as initial tuning protocol, and return to steady state of control under certain alarm conditions. The last, and most complex, method is to dynamically generate sequences. Each knowledge source transforms the state of the system from its initial state to its goal state. A sequence is recursively generated by comparing the desired goal and the current state with conditions of the knowledge sources.

The importance of deep modelling in intelligent supervisory control was addressed by Voss [1988]. It is noted that every intelligent autonomous system must have or develop a model of its world environment, and, in particular, have a model of itself, if it is to be able to manipulate its own environment. A deep model describing the physical shape, the structural and hierarchical component relations, the temporal behaviour, the functioning and the purpose of the real system is an essential source of information for most reasoning tasks concerned with diagnosis or control. Voss suggested that techniques of qualitative reasoning can be used for hierarchical modelling and deep-level knowledge representation.

In the approaches discussed above, the control functions are provided by conventional numerical techniques while AI methods are used in the hierarchical supervisory mode. However, there are other styles of approaches to system supervision, which adopt AI techniques in both supervisory level and primary control level. Abbod [1992] used a rule-based fuzzy logic controller to provide the primary control functions. The adaptive mechanism used in his approach consisted of two aspects. One was the technique of self-organising fuzzy logic control (SOFLC) [Procyk and Mamdani, 1979], which involves modifying the control rules via a performance index measure. The other was a self-tuning scheme based on fuzzy rules, which involves tuning the scaling factors for the fuzzy logic controller according to the performance monitoring result. The system supervisor selected the initial scaling factors and fuzzy rule base depending on the initial response of the process to an input control signal. A fuzzy-rule-based fault diagnosis mechanism was also developed to detect and diagnose via performance monitoring. In Handelman *et al* [1988], a neural network controller was employed in parallel with an expert rule-based controller. The expert controller and the neural network shared control, with the expert controller initially being in charge completely. As time passed, the neural network learned from the operation experience and took over the control gradually.

2.1.1 Observations on Intelligent Supervisory Control

All existent approaches to intelligent supervisory control can be seen as knowledge-based systems, although they use different architectures and inference methods. Generally, their architectures can be described by the block diagram shown in Fig. 2.1.

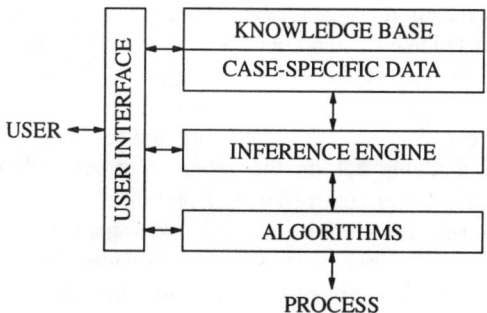

Fig. 2.1 Typical Architecture of Intelligent Control Systems

The algorithms consist of control algorithms, identification algorithms, and monitoring algorithms, which can be quantitative or qualitative. The control algorithms compute control signals according to the command given by the inference engine and measurement signals. The identification and monitoring algorithms monitor system behaviour and extract important information from the numerical signals, and then send the extracted information to the inference engine. The inference engine receives the information and applies the knowledge stored in the knowledge base to deduce a proper control strategy or tune the parameters of the control algorithms. In learning systems, the inference engine can also update the stored knowledge according to the observed system behaviour.

The heart of the intelligent control system is the knowledge base, which contains the problem-solving knowledge of the particular application. This knowledge can be represented in the form of if-then rules, graphs, objects, etc. The case-specific data includes the process specifications, performance criteria, reference inputs, and relevant information which cannot be inferred by the inference engine but is necessary for running a system. The user interacts with the intelligent control system through the user interface, which makes accesses more friendly and allows the user to maintain or modify the system.

The essential feature of this architecture, which makes intelligent control systems different from conventional ones, is the separation of the problem-solving knowledge, the inference engine, and algorithms. This allows the knowledge to be represented in a more natural form (e.g. if-then rules) rather than profound lower-level computer codes. Further, since the knowledge base is separated from the algorithms, it can be extended to involve the heuristic knowledge which the algorithms cannot provide. This heuristic knowledge supports higher-level reasoning and makes intelligent control systems more robust and flexible than conventional ones. Moreover, the separation of the knowledge base and control elements allows the inference engine and algorithms to be generalised for a variety of processes. In some learning systems, an intelligent controller can even begin to operate a process with an empty knowledge base and create a new knowledge base appropriate to the process [Abbod, 1992].

Currently, the success of intelligent supervisory control does not rely on the use of new mathematical methods but on the progress of AI techniques. In fact, the algorithms used in the lower control level of present supervisory systems are quite conventional. However, using heuristic knowledge to generate possible solutions for problems makes the supervisory systems intelligent. Here, knowledge representation plays a critical role in an efficient system, because, after a knowledge representation scheme has been decided, the corresponding inference engine is more or less easily implemented to apply the knowledge. A representation scheme must allow system builders to express the knowledge needed for a problem solution and result a computationally efficient program. The ability of an intelligent system to resolve problems also depends on how much knowledge can be articulated by the representation scheme.

The knowledge representation format most often adopted for intelligent control systems is the if-then rule. Although if-then rules are very good at representing heuristic knowledge, they also have several weaknesses: 1) they are not good for representing notions of time, causality and intent; 2) they do not provide the information to support reasoning about interactions between subsystems; 3) they are difficult to be verified and their correctness is difficult to be proved; and 4) since they are shallow, they can only tell "what" to do but cannot explain "why" to do so.

To overcome these difficulties, model-based approaches as in qualitative reasoning [Voss, 1988] were evolved to represent deep-level knowledge for intelligent supervisory control systems. A deep-model describes the behaviour of various components of a system and their functional relationships, so that causalities and interactions between components can be considered. Further, because a deep-model is built based on physical laws, it can be verified objectively. Qualitative causal graph models have been used by Leyval *et al* [1994] to represent deep-knowledge for a supervisory control system. Their implementation demonstrated that the supervisory system can give action advice and explanations via reasoning about a causal model.

Another difficulty in building intelligent control systems is knowledge acquisition. This difficulty is caused by the complexity of industrial processes. Structuring and representing complicated knowledge for efficient computation require a strong background in AI programming, such as making use of the high-level languages LISP and PROLOG, and object-oriented techniques. Accordingly, only professional AI programmers can cope adequately with these tasks. However, communicating correctly and completely the domain expert's knowledge to the AI programmers is difficult. At this point, developing methodologies which allow the domain expert to interact directly with the knowledge base to create, correct, and improve it becomes an important subject.

An approach to this subject is to develop an automatic knowledge base editor which can help domain expert build the knowledge base. An example is the program named Teiresias [Davis and Lenat, 1982] developed for the MYCIN expert adviser to help doctors add new rules to the system. Another way to accomplish this objective is to develop a system which can learn on its own to create a knowledge base through operating a process. Techniques of SOFLC and neural networks have been demonstrated successfully in this area.

Moreover, the approach of automatic deep model generation can also be used to resolve the problem of knowledge acquisition. An automatic modelling method based on qualitative bond graph reasoning has been developed by Linkens and Xia [1993]. Work in this book supercedes this approach and evolves a representation method to describe deep models. This representation allows deep models be described explicitly, and can be inserted easily into an intelligent control system without considering the details of the computer implementation, so that domain experts can use it to create a knowledge base by themselves. Here, the bond graph modelling language is employed to provide a systematic way for model building and representation, while qualitative reasoning is used as the basic strategy for reasoning about the deep-level knowledge stored in bond graph models. The inference engines and algorithms of controller design, auto-tuning, fault diagnosis, and planning are thus developed to utilise the knowledge represented in qualitative bond graphs.

2.2   QUALITATIVE REASONING

Newell and Simon [1976] have argued that intelligent activity, in either human or machine, is achieved through the use of: 1) symbol patterns to represent significant aspects of a problem domain; 2) operations on these patterns to generate potential solutions for problems; and 3) search to select a solution from among these possibilities [Luger and Stubblefield, 1989]. This argument makes explicit the underlying concept in qualitative reasoning research. The goal of qualitative reasoning is to develop symbolic computational theories to simulate the intelligence of engineers and scientists used in physical problem solving.

The program NEWTON built by de Kleer [1975] marked the beginning of qualitative reasoning. NEWTON was designed to solve problems concerning a single point mass sliding on a surface, based on algebraic techniques. In this program, the issues of what knowledge is required to solve classical physical problems and how to build a system that could solve them were investigated. Through the implementation of NEWTON, de Kleer concluded that qualitative reasoning is critical for comprehending the problem in the first place, formulating a plan for solving the problem, identifying which quantitative laws apply to the problem, and interpreting the results of quantitative analysis. In fact, the knowledge of physical kinematics and Newton's law are only a small fraction of the knowledge needed to solve problems. Most of the knowledge is "pre-physics", and considerable effort is required to codify it [de Kleer, 1993].

Disciplines of mathematics and physics were further relaxed in Hayes' Naive Physics Manifesto [Hayes, 1979]. Hayes proposed a formalism of the ordinary knowledge of physical common sense to reason about situations happening between events. The crucial difference between the naive physics and the mainstream approaches of qualitative reasoning is that the mainstream approaches represent knowledge in terms of physical laws rather than common sense. The standpoints taken in the mainstream approaches are that there should be no fundamental difference between "naive" physics and "ordinary" physics, and that knowledge with abstruse mathematical constructs can be represented explicitly and effectively.

1984 was a pivotal year for research in qualitative reasoning. In this year, the Journal *Artificial Intelligence* published a special volume entitled "Qualitative Reasoning about Physical Systems" [Bobrow, 1984], in which a number of significant contributions to the development of qualitative reasoning were brought together, thus promoting qualitative reasoning as an independent and established field in AI. Most papers in this special volume concentrated on the generation of qualitative causal explanations for time-evolved behaviour to explain how devices work. The best known of these researches, which provided a computational framework for qualitative analysis, include de Kleer and Brown's [1984] *device-centred* approach, Forbus'[1984] *process-centred* approach, and Kuipers'[1984] *constraint-centred* approach (improved as Kuipers [1986]).

In the *device-centred* approach, de Kleer and Brown presented algorithms for determining the behaviour of a composite device from laws governing the behaviour of its components, and introduced causality as an ontological commitment for explaining how devices behave. Three basic principles were established for performing the qualitative inference:

1)   *No-function-in-structure*: The laws of the components of a device may not presume the functioning of the whole. According to this principle, conditions which make the constitutive laws of a component recognised must be pre-defined in the structural description so that device's behaviour can be inferred reasonably.

2)   *Class-wide assumptions*: Those assumptions that are idiosyncratic to a particular device must be distinguished from those that are generic to the entire class of devices. According to the class-wide assumptions, the components of a device are idealised so that significant properties of them can be reasoned explicitly.

3)   *Locality*: The laws for a component cannot specifically refer to any other components. A component can only act on or be acted on by its immediate neighbours and its immediate neighbours must be identifiable a priori in the structure. According to this principle, causal interactions between components can be analysed.

Device components were connected via simple interactions, resembling the constitutive relationships of system dynamics. The laws governing the device were expressed in qualitative differential equations called confluences. Three qualitative values (+, 0, -) were assigned to confluences associated with each component. These qualitative values were used to represent physical quantities and could be manipulated through the qualitative operations shown in Table 2.1. The possible behaviour of a device can thus be predicted through the operations on the confluences of the device. The use of qualitative values and their operations has now become accepted as standard, and appears in most qualitative reasoning approaches.

$[X] + [Y]$:

[Y] \ [X]	-	0	+
-	-	-	
0	-	0	+
+		+	+

$[X] \times [Y]$:

[Y] \ [X]	-	0	+
-	+	0	-
0	0	0	0
+	-	0	+

Table 2.1 Qualitative Multiplications [de Kleer and Brown, 1984]

The *process-centred* approach was introduced by Forbus [1984]. This approach focused on reasoning about how things change in physical systems. Physical processes were seen as the mechanisms by which changes occurred. "Individual views" were adopted to model a set of active processes. An individual view consisted of four parts: 1) *individuals* — the objects existing in a process, 2) *quantity conditions* — comparisons between quantities of individuals or between an individual's quantity and a certain value (such as ZERO), 3) *preconditions* — the conditions must be true for a view, and 4) *relations* — physical laws among objects. Dynamics of a process were derived from the influences, which are qualitative equations indicating the directions of quantity changes in the process.

The *process-centred* approach is more suitable for reasoning about complex physical systems in comparison with the *device-centred* approach, because individual views can represent a large amount of knowledge for a complex system. For example, a thermo-fluid system contains a variety of materials and components. Interactions between these materials and components are very complex. Various assumptions are required to idealise the materials and components so that simple physical laws can be applied. The formation of individual views makes it possible to represent this knowledge explicitly. On the other hand, there are situations in which the use of a device-centred view is the more natural. An example would be the MOS circuit, where the components are almost ideal, and their properties and interactions are clear.

In contrast to the above two approaches, the *constraint-centred* approach does not aim to provide explanations for behaviour but rather directly simulates the behaviour of a physical system. This approach originated with Kuipers [1984] in the program QSIM. Qualitative simulation of a system started with a description of the known structure of the system, and an initial state; and then produced a directed graph consisting of the possible future states of the system and the "immediate successor" relation between states. The possible behaviours of the system were presented by the paths throughout the graph starting with the initial state. The structure of the system was described by a set of symbols representing the physical parameters of the system, and a set of constraint equations describing how these parameters may be related to each other [Kuipers, 1986]. Constraint equations were obtained from the qualitative abstraction of the ordinary differential equations governing the system.

Further requirements to improve the power of qualitative simulation have been stated by Kuipers [1993]. First, the causality employed in a system model was imposed from the outside by the user. However, there seem to be no tools currently available for describing and reasoning about causalities explicitly. A good representation for causalities is needed to support deriving system behaviours. Second, in QSIM certain conclusions which were intractable in the given model could be derived easily from an algebraically simplified abstraction of the model. Thus, an automatic general-purpose algebraic manipulation utility for model abstraction is necessary.

Approaches in qualitative reasoning since the first special volume have focused on formalising qualitative representations and reasoning techniques mathematically. Order of magnitude reasoning presented by Raiman [1986] aimed to develop a qualitative representation and reasoning methodology to cope with the ambiguity caused by the lack of quantitative information. Three relations between variables were given as:

1)   A Ne B — A is negligible in relation to B.
2)   A Vo B — A is close to B.
3)   A Co B — A has the same sign and order of magnitude as B.

Based on these relations, 30 inference rules were defined for relation interpretations, such as R_{21}: A Co B and C Co D $\rightarrow$ A·C Co B·D. However, no explicit definitions of "negligible" and "close" were available in this approach, and some rules appeared not true for some cases. For example, in R_{21}, if A = 1, B = 9, C = 100, and D = 900, the interpretation will become 1 Co 9 and 100 Co 900 $\rightarrow$ 100 Co 8100. The values 100 and 8100 are usually considered different orders of magnitude.

Mavrovouniotis and Stephanopoulos [1988] proposed a further detailed representation for the relations between variables (e.g. A is much smaller than B, A is slightly smaller than B, and so on) to try to resolve the interpretation problem. However, it is still difficult to explain explicitly the linguistic terms "much smaller", "slightly smaller", etc. At this point, fuzzy logic method seems to be an appropriate tool for representing the qualitative relations between variables, since it provides a formal methodology to reason about relationships using linguistic terms. Fuzzy logic and qualitative reasoning techniques have been integrated by Shen and Leitch [1990] to increase qualitative precision.

An approach motivated by improving the accuracy for process trend prediction was presented by Morgan [1988], and Cheung and Stephanopoulos [1990a, b]. In this approach, a qualitative state (QS) of a continuous variable (x) was represented by the qualitative vector:

$$QS(x,t) = \langle [x(t)], [\partial x(t)], [\partial \partial x(t)] \rangle,$$

with qualitative values drawn from the set $\{+, 0, -, ?\}$. Time-domain characteristics were expressed by a sequence of qualitative vectors. With this representation, underdamped, critical-damped, and overdamped responses of a system can be distinguished.

Development of reasoning techniques has been coupled to two active applications: fault diagnosis and automatic modelling. Approaches which contribute to fault diagnosis will be discussed in Chapter 6, while the significant research in automatic modelling are described below.

One reason which makes modelling difficult is that it requires identifying a large amount of domain knowledge pertinent to the individual application needs. An over-detailed model causes effort to be wasted during analysis, while an under-detailed model cannot adequately track the real system behaviour. Automatic modelling in qualitative reasoning attempts to develop algorithms to help the user abstract appropriate information from the domain knowledge, and to organise a model with as much detail as necessary to answer the questions of interest.

The most influential approach in this category is "compositional modelling" by Falkenhainer and Forbus [1991] based on Forbus' qualitative process theory. In compositional modelling, domain knowledge was decomposed into a number of pieces, called model fragments. Each model fragment represented a conceptually primitive phenomenon such as a physical process or one aspect of a component behaviour, and specified the functional interrelations of quantities in the phenomenon. A qualitative model was built through relating a minimal set of model fragments to provide the quantities of interest. The conceptual entities and relationships identified in qualitative analysis were used to guide the search for more detailed quantitative models. The combination of qualitative and quantitative models produces a self-explanatory simulator [Forbus and Falkenhainer, 1992], and accurate system behaviour can be provided by the numerical simulation on the quantitative model, while the explanations of the behaviour can be given by reasoning about the qualitative model.

There are two difficulties in the compositional modelling approach. One is that a system must be decomposed into several independent modules for model building. However, the system decomposition is not fixed, but depends on operating conditions and the purpose of analysis. An example is the decomposition of a heat exchanger in Falkenhainer [1992]. In this case, the quantity of interest is the temperature of the outlet plumbing of the heat exchanger, which is related to the variables of mass flow and heat transfer rates. However, the mass flow and heat transfer rates of the plumbing and heat exchanger are interdependent, so that these components must be seen as a module and cannot be decomposed. Determining the appropriate system decomposition requires detailed user understanding of the domain knowledge. The other difficulty is choosing the appropriate set of model fragments for a given problem. How a variable of a model fragment affects the quantity of interest cannot be known before the whole model is obtained. Also, it is difficult to decide whether or not the variable is an important factor to the quantity of interest without numerical details. How can a program provide a sufficient selection of relevant model fragments to best answer the given problem is an important issue for compositional modelling.

Another approach to automatic modelling builds models from system structures. Linkens *et al* [1991] developed a qualitative reasoning environment for automatic modelling and simulation of dynamic physical systems (QREMS), relating to the device-centred approach. In this approach, bond graphs were employed as a formal modelling language, where models of system components were represented by a set of physical primitives (e.g. resistance, capacitance, inertance, etc.) interacting by two interconnections (parallel junction and serial junction). Two variables, effort and flow, were associated with physical entities as primitive variables, on which qualitative reasoning about dynamic system was conducted (see further in Chapter 3). A model library has been built, which contains standard components (i.e. motor, transformer, etc.) in various detail.

A given system structure was decomposed into standard components and their connections. A bond graph model was thus constructed by interconnecting the simplest component models, and a set of qualitative equations relating the primitive variables of the components was then generated according to the bond graph model. Three constraints were used to investigate the correctness of the initial model:

1)   *Causal constraint*: An initial model must not contain any causal conflicts.
2)   *Zero-flow constraint*: When the output flow is set to zero, the states of all variables must be consistent with intuition and observation.
3)   *Zero-effort constraint*: When the output effort is set to zero, the states of all variables must not be against intuition and observation.

If the initial model contradicts any of the constraints, the components relevant to the contradictions will be re-modelled with their further detailed modes stored in the library. This investigating process is reiterated until the system model meets the constraints. Thus, a qualitative bond graph is obtained. When the values of system parameters are given, a bond graph toolkit ENPORT [Rosenberg, 1973] can be used to generate state-space equations of the bond graph model for numerical simulations.

The limitation of this approach is that it is difficult to choose the appropriate detailed-level of a model for a given problem. In this approach, a system component is composed of several physical primitives. However, the importance of a primitive for the problem of interest cannot be ensured because of the lack of numerical details. A simple model which can meet the above three constraints may not be detailed enough to answer the problem of interest. On the other hand, building a model at the most detailed level guarantees the completeness of the model, but may also result in an unnecessarily large model. Thus, a model abstraction method, which can determine the insignificant or the necessary primitives according the problem of interest, is required. A general model abstraction method is one of the most important tasks in improving the ability of qualitative bond graph modelling.

This section introduced the basic concept of qualitative reasoning. A more detailed survey can be seen in [Werthner, 1994].

2.2.1   Observations on Qualitative Reasoning

As discussed in Section 2.1, knowledge representation plays a critical role in intelligent supervisory control systems. How intelligent a system is depends on how much knowledge can be represented clearly and how deep is the knowledge which can be utilised for problem-solving. Qualitative reasoning here shows strong power in deep-knowledge representation and utilisation. Automatic modelling provides a systematic way for knowledge acquisition. A number of knowledge representation formations have been developed along with their utilisation methods, e.g. device-centred, process-centred, and constraint-centred approaches. Efforts on qualitative reasoning have provided a solid foundation for building intelligent supervisory control systems.

Qualitative reasoning has two problems which have never appeared in quantitative approaches. One is that qualitative reasoning may produce results which are ambiguous. Although de Kleer and Brown [1986] have claimed that ambiguity provides the advantage that all possible behaviours of a physical system can be analysed, it really gives problems when a certain answer is required. Techniques such as transition analysis and feedback analysis [Williams, 1990] have been developed to resolve ambiguities, but these techniques do not resolve all types of ambiguities (such as simultaneity) and, thus, quantitative information is needed.

Another difficulty in qualitative reasoning is the representation of "time". In reasoning about the dynamic behaviour of systems, it is necessary to consider the way in which time is represented. Most qualitative reasoning approaches consider time to be broken into a succession of discrete periods. In each time period, all particular conditions hold true. In other words, specified qualitative values are assumed to be unchanged over a time period. Of course, this assumption is not true in the real world. History-based representation [Hayes, 1985; Williams, 1986] has thus been developed to resolve this problem, whereby a variable is represented by its qualitative value along with the time interval at which the value happened. However, it is difficult to track changes of every variable separately in a real-time system (it needs a large number of sensors). Also, it is difficult (or impossible) to predict how fast a variable will change with the use of qualitative information.

Shen and Leitch [1993] utilised fuzzy set techniques to represent time for qualitative simulation. In this approach, the fuzzy qualitative state of a system variable is described by a pair $<A, B>$, where A and B are fuzzy quantities. A denotes a interval-valued magnitude and B represents the fuzzy rate of change of the variable. Thus, the persistence time (ΔT_p) of the variable can be inherently determined by the extent of the fuzzy magnitude and the rate of change of the state, where ΔT_p records the amount of time a variable remains within a given qualitative state (A). However, since the persistence time is obtained from a fuzzy state, it provides only the possible range of time. As a result, present qualitative simulations can predict "what" will happen and explain "why" it will happen, but cannot tell exactly "when" it will happen.

Due to these problems, qualitative reasoning is more suitable for off-line applications (i.e. simulation and computer-based tutoring) or cause-effect explanation (i.e. fault diagnosis) than applications which require certain answers and precise time representation such as predictive control. This book will not attempt to resolve these problems, but try to prevent them and develop on-line applications for system supervision.

2.3   BOND GRAPHS

A bond graph is a system representation of power interactions with connecting lines, "bonds," which carry both power variables and causalities between power variables. The bond graph technique, used for modelling dynamic multiport systems, was created in 1959 by Paynter [1959]. This approach was motivated by generalising the electric circuit diagram concept to develop a general theory for engineering systems. Paynter suggested that energy and power are the fundamental dynamic variables which govern all physical interactions and transactions. Following the original idea, Rosenberg and Karnopp [1972] proposed the theoretical basis and definitions of this method. From then on, bond graphs have become a formal modelling language of dynamic systems.

A procedure for the systematic generation of linear state equations in terms of energy variables was proposed by Rosenberg [1971]. Martens [1973] extended the equation formulation method to cover non-linear systems. The notion of causality was introduced by Karnopp [1975] to guide the state equation formulation. A fast complete method for automatically assigning causality to bond graphs has been developed by Hood *et al* [1989]. How bond graphs and the sequential causality assignment algorithm (SCAP) can be used in building a qualitative model of a system was presented by Barreto [1988]. Barreto pointed out that a qualitative model is constituted by facts and rules. However, it is not easy to choose the cause and effect of a functioning rule in each case. Bond graphs provide a systematic causality assignment method which can be used to resolve the causal ordering problem for qualitative modelling. Causality in both qualitative reasoning and bond graphs will be discussed further in the next section.

Applications of the bond graph technique were made initially in mechanical, electrical, and thermodynamic systems. In recent years, the applications have been extended further to chemical, fluidic, biological, economic, and even agricultural systems. A number of computer-aided modelling and simulation programs based on bond graphs, such as ENPORT [Rosenberg, 1973], CAMP (*Computer-Aided Modelling Program*) [Granda, 1985], and CAMAS (*Computer-Aided Modelling, Analysis and Simulation*) [Broenink and Nijen Twilhaar, 1985], have been developed widely through scientific and engineering fields (see a survey in [Filippo *et al*, 1991]). The variety of systems for which bond graphs have been employed demonstrates that the bond graph technique is a useful and versatile tool for modelling and simulation. This is due to the fact that the simple definition of the basic variables of a system facilitates the use of bond graph theory for modelling many types of systems.

2.4   CAUSALITY

To organise component constitutive laws into sets of equations, one needs to make a series of cause-effect decisions to combine the algebraic relations between variables. In building models, scientists and engineers usually assign causalities intuitively according to their knowledge of physical phenomena. For example, we usually view the causalities in Ohm's law (V = IR) in the directions of "V causally determines I" or "I causally determines V" rather than "V and I causally determine R". The causality assignment in this example seems quite easy because we are very familiar with the Ohm's law. However, assigning causalities for complex systems requires cross-domain knowledge and the analysis of composite functional relations, which make modelling difficult. To resolve this problem, several domain-independent causality assignment methods have been developed independently in the approaches of qualitative reasoning [Iwasaki and Simon, 1986; de Kleer and Brown, 1986] and bond graphs [Rosenberg and Karnopp, 1983]. Currently, causality assignment is still a controversial issue in qualitative reasoning, but, on the other hand, a formal method for causality assignment based on bond graphs has already been built and computerised. As suggested by Barreto [1988], introducing bond graph techniques into qualitative reasoning seems a good way to resolve the causality assignment problem in qualitative reasoning.

In bond graphs, "bonds" play the role of representing the interactions between variable pairs (efforts and flows) and their causalities. Fig. 2.2 shows two components, A and B, bonded together with the short perpendicular line (the causal stroke) on the end of the bond near A. This means that B determines the effort $e(t)$ and impresses it on A, which causes A to respond with the flow $f(t)$, which is returned to B. Then, the flow causes B to respond with an effort. It can be seen as B pushing the effort on A and A pointing the flow at B. Thus, the cause-effect relations for efforts and flows are represented in opposite directions.

Fig. 2.2 The Representation of a Bond and its Causal Stroke

Usually, a bond is further represented with a half arrow on one end of the bond to indicate the direction of power delivery. The assignments of the half arrow and the causal stroke are completely independent, so there are four possible representations of a bond:

$$A \longmapsto B \quad A \mathrel{\mapsfrom} B \quad A \longrightarrow B \quad A \mathrel{\longleftarrow} B,$$

in which the half arrows describe the positive directions for efforts and flows, and the causal strokes indicate the input-output relations for efforts and flows.

Power sources have only one causality, since they are assumed to apply a predetermined effort or flow on other components. For example, an effort source always outputs an effort, so the only possible causality is

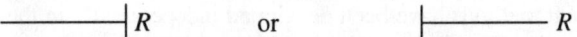

It also makes sense that resistance should take whatever causality it is given. For example, given a voltage, a resister will respond with a current; while given a current, a resister will respond with a voltage. Thus, causality of a resistance can be

$$\longmapsto R \qquad \text{or} \qquad \vdash\!\!\!\longrightarrow R.$$

Similarly, capacitance and inertance can have two causalities. For a capacitor, given an input current, it will respond with an output voltage. For an inductor, given an input voltage, it will respond with an output current. This kind of causality is called *integral causality*, since the input is integrated to produce the output. The integral causal patterns are

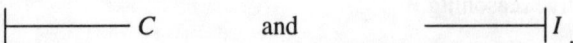

Reversing the integral causality leads to the *derivative causality*, where the causality between the input and output has a derivative direction. The derivative causal patterns are

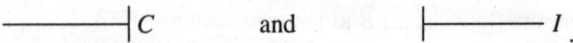

Although integral and derivative causalities both exist in the real world, the integral causality is preferred in bond graph modelling. This is because the mathematical representation of integral causality makes it easy to generate state-space equations. If all the capacitances and inertances of a system can be assigned with the integral causality, all the state variables of the system can be independent. However, the nature of a particular system may force a capacitance or inertance to have an undesired derivative causality, which means that some state variables are algebraically coupled and cannot be independent. In this situation, further algebraic operations are required for formulating explicit state-space equations, which make the equation formulation difficult [Rosenberg and Karnopp, 1983].

It is interesting that researchers in qualitative reasoning have reached a very similar conclusion through analysing physical phenomena that the causality always flows along the direction of integration [Williams, 1990]. An example, *RC* high pass filter, used by Williams for causality analysis is shown in Fig. 2.3.

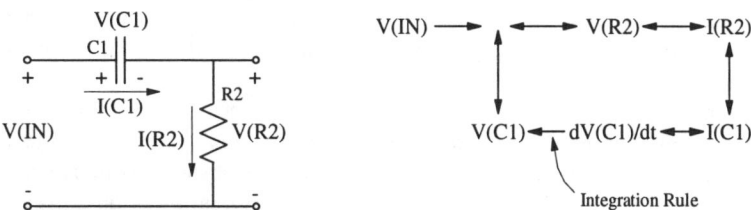

Fig. 2.3 *RC* High Pass Filter and its Causal Path [Williams, 1990]

The causal path in this case was analysed as following: When the input voltage begins to rise the capacitor initially acts like a "battery", thus the change in the input voltage is directly transmitted to the resistor's voltage. This produces a current through the resistor which charges the capacitor and causes V(C1) to increase. The causality here flows through the integral direction from dV(C1)/dt to V(C1).

On the other hand, the *RC* high pass filter can be modelled using bond graphs. The bond graph model and its causality assignment are shown in Fig. 2.4, where the symbol "1" means that the effort source, resistor, and capacitor are related with a serial interaction. The causality is assigned based on the principle that integral causality is preferred. The detailed methodology for building bond graph models will be discussed in Chapter 3.

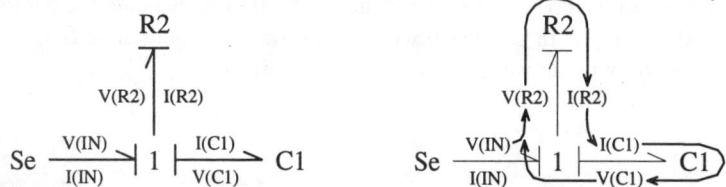

Fig. 2.4 The Bond Graph Model of *RC* High Pass Filter and its Causal Path

According to the causality assignment, the causal path can be determined as shown in Fig. 2.4. This causal path is exactly the same as that described by Williams, but it is obtained from a formal causality assignment procedure rather than an informal analysis. This case shows that bond graphs can be used as a formal modelling language for qualitative reasoning to resolve the problem of causality assignment.

2.5   DISCUSSION

As seen from the above example, bond graphs can also provide a qualitative system representation, so they can be employed as the formal modelling language for qualitative reasoning. Bond graphs become quantitative only when the parameters are added. If it is necessary, system parameters can be inserted into a qualitative bond graph model to improve its precision. Further, bond graphs provide a systematic way for building deep models, in which the system structure and its interactions can be represented explicitly. On the other hand, qualitative reasoning provides formal reasoning methods to reason about the information carried by the bond graph models. Their integration performs a powerful methodology to represent and reason about deep-level knowledge, around which an intelligent supervisory control system can be built.

Although the causality assignment problem can be resolved by using bond graphs, this book does not utilise the technique of causality assignment. In fact, the applications developed in this book do not employ the notion of causalities. There are three reasons for this. First, bond graph models can be built in terms of system structures without taking account of causalities. This will make model building much easier and avoid the inference to be disturbed by the undesired derivative causality. Second, the causality in bond graphs is assigned to formulate state-space equations. However, the qualitative representation used in this book can be formulated without the consideration of causalities. Third, after a model has been built, problems can be resolved usually without taking account of causalities. For example, in most cases, the causality of "V and I causally determine R" is violated in Ohm's law, but we do often use the values of V and I to evaluate the parameter R. The method developed in Chapter 6 will demonstrate that qualitative fault diagnosis can be successful without using the notion of causalities.

CHAPTER 3

QUALITATIVE
BOND GRAPH MODELLING

3.1 INTRODUCTION

Conventional engineering approaches, along with quantitative mathematics have been under development for several centuries and have been proved to be very sophisticated and successful in systematically analysing the physical world. However, there are still some circumstances in which quantitative approaches are not easy to apply; for example, where it is difficult to obtain precise information about the values of system parameters, or where the system is not in a conveniently analytic form, or where there are real constraints on memory size and time for complex calculations. In these circumstances, qualitative representations appear very useful because they possess an ability to reason with incomplete and weak numerical information. A further important reason for developing qualitative physics is that engineers do not reason exclusively about a system in terms of the precise values and interrelationships between parameters but rather reason about these values and interrelationships at a more abstract and qualitative level [Williams, 1991].

This chapter will consider qualitative representations. Both the strengths and the difficulties of applying qualitative reasoning to system supervision will be analysed. Then, an integrated modelling method combining qualitative reasoning with bond graphs will be proposed in order to overcome some difficulties accompanied with the ambiguous properties of qualitative reasoning and to resolve modelling problems caused by the complexity of engineering systems. This modelling scheme can generate formulated qualitative models which are applicable to mechanical, electronic, electrical, thermal, or fluid physical systems, or indeed to systems comprising several energy forms. Further discussions about describing dynamic systems using qualitative bond graph models for feedback control and fault diagnosis will be given in Chapters 4 and 6.

3.2   OBSERVATIONS ON QUALITATIVE REPRESENTATIONS

There are two major approaches which make use of qualitative representations for engineering problem solving: *qualitative reasoning* and *fuzzy set theory*. Some major research in qualitative reasoning proposed by de Kleer, Brown, Forbus, Kuipers, etc. has been discussed in Chapter 2, where the behaviour of a physical system is represented in terms of values drawn from the set {+, 0, -, ?} where "+" and "-" denote two partitions in a measurement range, "0" denotes the boundary between them, and "?" describes an ambiguous value. In contrast, fuzzy set theory applies linguistic variables (such as "positive big", "positive small", "negative medium" and so on) to qualitative value representations to simulate human thinking and the inexact nature of the real world [Zadeh, 1987]. A fuzzy set is defined as follows.

Let U be a collection of objects denoted generically by $\{u\}$, where U is called the universe of discourse and u represents the generic element of U. A fuzzy set A in a universe of discourse U is characterised by a membership function μ_A which is usually defined in the following form:

$$\mu_A : U \rightarrow [0, 1],$$
$$u \longmapsto \mu_A(u),$$

where [0, 1] denotes the interval of real numbers from 0 to 1, inclusive.

Then, the fuzzy set A can be written concisely as

$$A = \int_U \mu A(u) / u.$$

A value $\mu_A(u_i)$ assigned to the element u_i of the universal set U falls within a specified range and indicates the membership grade of this element in the set. A larger value denotes a higher degree of set membership. With this formulated form, approximate commonsense and human knowledge can be generally represented and applied to computers to cope with complex problems. Here, only the basic concept of fuzzy set theory has been addressed for comparison purposes. Detailed discussions can be found in Zadeh [1965], Kaufmann [1975, 1985] and Klir [1988].

3.2.1   Comparison between Qualitative Reasoning and Fuzzy Set Theory

The fundamental difference between qualitative reasoning and fuzzy set theory is motivation. Qualitative reasoning has been developed to be a computational theory of the core skills underlying engineers and scientists [Williams, 1991]; it is concerned with representing and reasoning about the physical world. Accordingly, its concepts are to hypothesise, predict, control, and diagnose physical mechanisms. On the other hand, the goal of the fuzzy set theory is to build a general mathematical formulation to cope with complexity and uncertainty of knowledge about the real world. The aim of this theory is to capture not only knowledge about physical tasks but also knowledge of non-physical problems. Hence, fuzzy set theory can be more widely applied to the fields of psychology, engineering, economics, medicine, sociology, and meteorology [Gupta, 1985; Klir, 1988; Wang, 1983]. This fundamental variance indicates two further important differences between these two theories.

The first difference is the knowledge representation method. In qualitative reasoning, a physical system is described by existing theories of physics (e.g. circuit theory [Williams 1984], physical system dynamics [de Kleer, 1984], kinematics [Forbus, 1987], and thermodynamics [Skorstad, 1989]) and mathematics (e.g. differential equations [Kuipers, 1986] and algebra [Struss, 1990]) with qualitative formalisms. On the other hand, in fuzzy set theory, an object is characterised by a membership function, while assignment of grades of the membership function to the elements of the object is based on statistical considerations [Murthy, 1990; Kruse, 1982; Kwakernaak, 1978, 1979]. For example, representation of a mass-spring-friction system (Fig. 3.1), using the method of qualitative reasoning proposed by de Kleer [1984] will be as follows:

The behaviour of the mass is described by Newton's Law $F = ma$ or qualitatively $[F] = \partial v$, where [ ] denotes a qualitative value +, −, 0, or ?. Hooke's Law for the spring $F = -kx$ becomes $\partial F = -[v]$. The resistance of the shock absorber is modelled by $[F] = -[v]$ and $\partial F = -\partial v$. For simplicity sake, define $x = 0$ as the mass position with the spring at equilibrium, and $x > 0$ to be to the right as shown in Fig. 3.1. The net force on the mass is provided by the spring and shock absorber: $Fmass = Fspring + Ffriction$ or qualitatively $[Fmass] = [Fspring] + [Ffriction]$. Besides, according to the physical laws, a set of equations is derived to govern the movements of this dynamic system: $\partial v = [Fmass]$, $\partial Fspring = -[v]$, $\partial Ffriction = -\partial v$ $= -[Fmass]$, and $\partial Fmass = \partial Fspring + \partial Ffriction = [Fmass] - [v]$. Then, the possible behaviour of the system is illustrated in Table 3.1 via qualitative inference based on the governor equations [de Kleer and Bobrow, 1984].

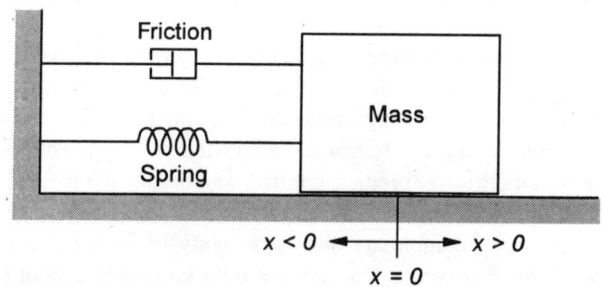

Fig 3.1 Mass-Spring-Friction System

	1	2	3	4	5	6	7	8	9	10	11	12	13
$[F_{mass}]=$	0	–	–	–	0	+	+	+	+	+	0	–	–
$[F_{friction}]=$	0	–	0	+	+	+	+	+	0	–	–	–	–
$[F_{spring}]=$	0	–	–	–	–	–	0	+	+	+	+	+	0
$[v]=$	0	+	0	–	–	–	–	–	0	+	+	+	+
$\partial F_{mass}=$	0	?	+	+	+	?	?	?	–	–	–	?	?
$\partial F_{friction}=$	0	+	+	+	0	–	–	–	–	–	0	+	+
$\partial F_{spring}=$	0	–	0	+	+	+	+	+	0	–	–	–	–
$\partial v=$	0	–	–	–	0	+	+	+	+	+	0	–	–

Table 3.1 Qualitative Representation of Mass-Spring-Friction System
(from de Kleer and Bobrow, 1984)

Alternatively, if the system is to be represented via the fuzzy set theory, then the first task is to collect some characteristic variables for the concerned system behaviour and then define the values of these variables with linguistic descriptors. In this case, for instance, the characteristic variables are found to be position, velocity and acceleration of the mass (x, v, and ∂v); where the linguistic values of x and ∂v may be "positive small (PS)", "positive big (PB)", "negative small (NS)", and "negative big (NB)' and the values of v may be "positive slow (PS), "positive fast (PF)", "negative slow (NS)", and "negative fast (NF). Therefore, the system behaviour will be represented in the form of natural language as shown in Table 3.2.

If x is PS and if v is PF, then ∂v is NS.
If x is PS and if v is NF, then ∂v is NS.
If x is PB and if v is PS, then ∂v is NB.

If x is PB and if v is NS, then ∂v is NB.

If x is NS and if v is PF, then ∂v is PS.

If x is NS and if v is NF, then ∂v is PS.

If x is NB and if v is PS, then ∂v is PB.

If x is NB and if v is NS, then ∂v is PB.

Table 3.2 Fuzzy Representation of Mass-Spring-Friction System

Fuzziness in this representation is caused by the vague boundaries of the linguistic values. In contrast with a crisp set, a fuzzy set defined in a given universe of discourse has no sharp distinction between the members and non-members of the class described by the fuzzy set. In this case, the meaning of "fast" could be very different for individual people. Furthermore, "fast" itself is also a vague concept in the human mind. In order to formulate this fuzzy representation, a set of possible membership functions is defined mathematically for the distance, velocity, and acceleration via experience. For example, velocity close to "positive fast" could be:

$$\mu_{PF}(v) = \begin{cases} 1 & \text{, when } 1\,\text{m/sec} \leq v, \\ \dfrac{v - 0.4}{0.6} & \text{, when } 0.4\,\text{m/sec} \leq v < 1\,\text{m/sec}, \\ 0 & \text{, when } v < 0.4\,\text{m/sec}. \end{cases}$$

With this method, a complete family of membership functions representing the set of values of x, v, and ∂v can be stated. Naturally, intermediate linguistic values (i.e. "positive medium" and "negative medium" etc.) used in the fuzzy sets will produce more detailed description of the system behaviour.

In essence, the knowledge representation of qualitative reasoning is objective, since it is based on objective physical laws; whereas, representation in the fuzzy set theory is subjective because of the subjective nature of membership grading. Accordingly, understanding the physical structure of a system and finding its ruling physical laws are necessary for qualitative reasoning descriptions. Instead, long-term experience about a system and finding an appropriate universe of discourse and membership functions for individual variables are the essential parts of fuzzy representations.

The second difference between these two approaches is that although qualitative values used in qualitative reasoning are less precisely measured than numerical values, they are crisp and distinct from the type of imprecision represented in fuzzy set theory [Morgan, 1988]. In qualitative reasoning, all values are either on the side of +, or on the other side of -, or exactly on their boundary 0. Undetermined values are represented by the symbol ? rather than fuzzy values. However, in the fuzzy set theory, an uncertain value can be determined in terms of possibility theory and represented with its grade of membership in a fuzzy set [Klir, 1988].

3.2.2   Why Qualitative Reasoning?

As a basic strategy for supervisory control systems, qualitative reasoning has many advantages over the forms of reasoning performed by the traditional mathematical methods or fuzzy set theory. Firstly, qualitative reasoning can be employed to build a deep-level knowledge model to represent the relationship between system structure and behaviour. In a deep-level knowledge model, structural information, referring to system components and their connections, is described in qualitative forms, while system behaviour is derivable from component behaviour and connection constraints. Therefore, internal mechanisms of a system and cause-effect relationships of system behaviour can be analysed via reasoning with qualitative models. This capability is significant for intelligent system supervision.

Secondly, qualitative reasoning represents variable values with simple {+, 0, -} subdivisions. Making use of this qualitative scale removes some potential difficulties relating to units of system variables, since all values are either on one side of a boundary or the other (or exactly on the boundary) [Morgan, 1988]. Thus, relations between variables can be expressed easily without taking their units into account. Furthermore, relations between subsystems are relatively easy to obtain by directly connecting their qualitative models and do not depend on the system complexity [Francis, 1984]. These benefits facilitate the applicability of supervisory methods to more complex engineering systems.

Finally, qualitative reasoning supports high-level reasoning in both its representation and inference. It describes a situation in terms of gross features, insightful concepts and distinctive states without using numerical details [Xia, 1991]. Therefore, a system can be represented and analysed when its low-level details are unknown. However, when numerical details are necessary for the purposes of accuracy, recognised numerical information can be inserted into qualitative models to improve their accuracy (an example will be discussed in Chapter 4). Consequently, qualitative reasoning provides supervisory control systems with both inference ability and accurate performance when necessary.

3.2.3   Problems to be Resolved

Applying qualitative reasoning to system supervision tasks requires building effective models to support both high-level conceptual inference and low-level primitive control. An effective qualitative model should possess the following capabilities:

- Qualitative models should describe explicitly the locations of system components and their interconnections whereby interaction between the parts can be analysed. Therefore, for fault diagnosis aspects, an inference mechanism can reason about fault locations via this structural information.

- Qualitative models should describe simply the relations between system input and output so that a primitive control algorithm can directly adjust the system input according to the output error rather than reason about every internal state of system components. It not only enhances computational efficiency but also avoids problems of ambiguity arising from qualitative operations.

- Qualitative models should be applicable across domains to represent complex dynamic systems with a uniform format and should be systematically built by using a formal modelling language. Thus, qualitative supervisory control methodology can be widely applied to cope with complex engineering systems.

- Qualitative models should allow the integration of qualitative representations and quantitative information without using unnatural descriptions so that the strengths of each can be utilised.

- Qualitative models should be established in a form which will overcome the difficulty of higher-order qualitative derivatives (introduced by de Kleer and Bobrow [1984]) to improve the accuracy of conceptual inference.

The remaining section of this chapter proposes an integrated qualitative bond graph modelling method to meet the above requirements.

3.3   BOND GRAPHS

Bond graph methodology provides a formal and systematic language for modelling dynamic systems. It incorporates physical assumptions and issues made about system models in an explicit and precise way. In practical engineering, bond graphs are typically used to help generate the differential or the state equations of systems for standard numerical simulation. However, they can also be used for various forms of qualitative reasoning [Top, 1991] which made no reference to system differential equations at all.

The bond graph modelling language is composed of a highly organised domain independent syntax. It makes use of a small number of primitive elements to describe a limited set of physical concepts which can be employed as a set of general conventions to represent physical models. The primitive entities are: three passive elements, resistance (R), capacitance (C), and inertial elements (I); two distribution elements, transformer (*TF*) and gyrator (*GY*); and two ideal sources, effort source (S_e) and flow source (Sf). The interactions among these elements are expressed in terms of their energy transformations. Here, directions of energy flow are indicated by directed bonds, each of which has two associated power variables — effort (*e*) and flow (*f*). Connections between multiple elements are established by two junctions — serial junction (*1-junction*) and parallel junction (*0-junction*). Functions and descriptions of the primitives used in this book are discussed as follows.

Bonds and Power Variables

In bond graph language, a bond marks a power connection between two parts (*A* and *B*) of a system. It is represented by a line with a half arrow which indicates the direction of the actual power flow:

$$A \xrightarrow[f]{e} B.$$

The symbols of its associated power variables effort (*e*) and flow (*f*) are separately written above and below the bond line. Effort and flow are generalised variables where effort stands for force, pressure, torque, voltage and absolute temperature, etc. and flow stands for velocity, flow quantity, rotation frequency, current and entropy flow, etc. In addition, the product *ef* is the power on the bond, so effort and flow are called power variables. It should be noted that the half arrow always indicates the direction of power flow when the product *ef* is positive.

In this book, the power variables will not be shown in bond graphs except by a series of numbers to label the bonds.

Passive Elements

There are three fundamental ideal passive elements in physical systems: resistance (*R*) representing power dissipation, capacitance (*C*) and the inertial element (*I*) describing two forms of energy storage. Since they contain no sources of power, we call them passive elements.

The first considered element is resistance which represents damper, absorber, resistor, fluid resistance or heat resistance in different domains. Its bond graph symbol is shown as:

$$\longrightarrow R.$$

The effort (e) and flow (f) on the bond of an R element are directly related by a constitutive law (i.e. in electrical systems, it is Ohm's law)

$$e = Rf. \tag{3-1}$$

Since resistance always dissipates power, the half arrow on the bond of an R element always points towards R.

The next considered element is capacitance (C) which represents components such as a condenser, accumulator, linear spring and torsional spring, etc. Its symbol is shown as:

$$\longrightarrow C.$$

The relationship between its power variables, e and f, obeys Hooke's law and is written as:

$$f = C\frac{de}{dt}, \tag{3-2}$$

$$e = \frac{1}{C}\int f dt. \tag{3-3}$$

The half arrow pointing toward C means that positive ef represents power flow to the capacitance. Here, the power flow indicates the rate of energy storage. Yet, when energy is released from a C element, the value of ef will be negative.

The last element to be considered is an inertial element (I) which is used to represent the behaviour of a mass, inductor and flywheel, etc. Its bond graph description is shown as:

$$\longrightarrow I.$$

The behaviour of an I element is governed by Newton's law and is written as:

$$e = I\frac{df}{dt}, \tag{3-4}$$

$$f = \frac{1}{I}\int e dt. \tag{3-5}$$

The I element is energy-conservative, just like the C element. The half arrow again indicates that whenever ef is positive, energy is flowing into I and is stored by the inertial element.

Distribution Elements

Engineering systems have two types of elements which transfer energy between two parts of a system. In bond graphs, they are generally represented as transformers (TF) and gyrators (GY).

A transformer (such as a lever, electric transformer, pump, gear reducer, etc.) has two ports, with the efforts at the two ports being proportional to each other, as are the flows. Its symbol is

$$\xrightarrow{\quad 1 \quad} TF \xrightarrow{\quad 2 \quad}$$
$$\text{b/a}$$

where the b/a is a proportionality factor called the modulus of the transformer. Thus, the behaviour of the transformer can be represented by the following generalised notions:

$$e_1 = \frac{b}{a} e_2, \qquad\qquad (3\text{-}6)$$

$$\frac{b}{a} f_1 = f_2. \qquad\qquad (3\text{-}7)$$

The power relation between two ends of a TF element will then be derived by multiplying Eqs. (3-6) and (3-7):

$$e_1 f_1 = e_2 f_2, \qquad\qquad (3\text{-}8)$$

which indicates that power flowing into port 1 is always equal to power flowing out of port 2.

A gyrator (such as an electric motor or a hydrostatic motor, etc.) relates the effort at one port to the flow at the other port, and is symbolised by:

$$\xrightarrow{\quad 1 \quad} GY \xrightarrow{\quad 2 \quad}$$
$$\text{b/a}$$

The modulus b/a enters the gyrator constitutive laws as follows:

$$e_1 = \frac{b}{a} f_2, \tag{3-9}$$

$$\frac{b}{a} f_1 = e_2. \tag{3-10}$$

As in the case of the transformer, power is conserved.

Power Sources

Corresponding to the power variables effort and flow, bond graphs employ two typical ideal power sources, effort sources (S_e) and flow sources (S_f), to represent active elements of systems. Their symbols are:

$$S_e \longrightarrow, \qquad\qquad S_f \longrightarrow.$$

The ideal sources supply any amount of power to whatever they are connected. An effort source supplies functional effort without depending on its flow, while a flow source supplies functional flow without depending on its effort. The half arrow pointing from a power source denotes that power is being supplied by the source when its *ef* is positive.

Connections

Bond graphs contain two types of connections, serial junction (*1*) and parallel junction (*0*), to connect several bonds in a bond graph model. Junctions are power conserving at each instant and the power transports of all bonds add to zero at all times. The symbols of a 3-port 1-junction and a 3-port 0-junction are shown as follows (of course, they could be 4-port or more):

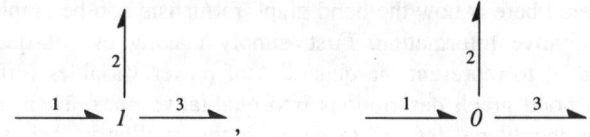

The 1-junction can also be called a common-flow junction which has the following properties:

- There is a single flow variable common to all bonds connected to a 1-junction.

- An algebraic sum of all efforts on the bonds attached to a 1-junction is zero, where the signs in this algebraic sum are determined by their half arrow directions.

- The net power flowing to a 1-junction on all its bonds is zero at any instant of time.

As a result, the behaviour of a 1-junction can be represented generally by the following forms:

$$e_{in1} + e_{in2} + \cdots + e_{inm} = e_{out1} + e_{out2} + \cdots + e_{outn}, \qquad (3\text{-}11)$$

$$f_{in1} = f_{in2} = \cdots = f_{inm} = f_{out1} = f_{out2} = \cdots = f_{outn}. \qquad (3\text{-}12)$$

The 0-junction is also called a common-effort junction. It is just the bond dual of the 1-junction so that the roles of effort and flow are reversed, and its mathematical representation is:

$$e_{in1} = e_{in2} = \cdots = e_{inm} = e_{out1} = e_{out2} = \cdots = e_{outn}, \qquad (3\text{-}13)$$

$$f_{in1} + f_{in2} + \cdots + f_{inm} = f_{out1} + f_{out2} + \cdots + f_{outn}. \qquad (3\text{-}14)$$

The basic forms of the elements used in this research to represent all types of physical systems have now been discussed. More detailed discussion can be found in [Rosenberg, 1983].

3.4   QUALITATIVE BOND GRAPH MODELLING

The problem considered here is how the bond graph formalism can be employed in order to obtain qualitative information. First, simply making use of qualitative values (i.e. +, 0, and −) to represent the quantities of power variables (effort and flow) will transform bond graph descriptions into qualitative ones. Then, a set of qualitative operators should be defined to describe the qualitative behaviour of physical elements in a formal mathematical form. The basic possible operators for qualitative values are generally drawn from the standard operators for real numbers and have been defined in many papers as has been discussed in Chapter 2. Finally, system behaviour will be described by connecting all the qualitative representations of the system elements by 1- or 0-junctions.

3.4.1   Qualitative Representation in Bond Graphs

In this book, qualitative values of effort and flow are represented by capitals E and F, while qualitative operators are drawn from the set $\{+, -, \times, /, =\}$ which have the same meanings as their corresponding operators for real numbers. The parameters of resistance, capacitance and inertial element are described respectively by symbols R, C and I. Qualitative representations of bond graph primitives are established with these symbols.

Passive elements

Obviously, the qualitative notation about a resistance can be directly converted from its quantitative form Eq. (3-1):

$$E = R \times F. \tag{3-15}$$

However, the qualitative representations of the power storage elements C and I are different from their quantitative formulas. The reason is that the supervisory control method developed here is computerised, so the qualitative representation should be discrete. Accordingly, we do not take integration and differentiation as qualitative operators.

Let us consider the qualitative representation of a C element. Firstly, its quantitative behaviour can be approximated as follows:

$$f(t_2) = C\frac{de(t_2)}{dt} \approx C\frac{(e(t_2) - e(t_1))}{\Delta t}, \tag{3-16}$$

where $t_1 < t_2$ and $\Delta t = t_2 - t_1$. Then, the qualitative behaviour of a C element can be written as:

$$F(t_2) = C \times \frac{(E(t_2) - E(t_1))}{\Delta t}. \tag{3-17}$$

In a discrete system, Δt can be seen as a sampling period, while the value of Δt is always positive. As a result, it has no influence on qualitative behaviour and can be neglected. Thus, the qualitative representation of a C element is finally produced:

$$F(t_2) = C \times (E(t_2) - E(t_1)). \tag{3-18}$$

This formula will lead to some error when representing behaviour around an inflection point, but it can still correctly relate power variables to describe the global behaviour and function of C elements.

Similarly, the integral equation Eq. (3-3) can be converted to the qualitative form:

$$E(t_n) = \frac{1}{C} \times (F(t_n) + F(t_{n-1}) + F(t_{n-2}) + \ldots), \tag{3-19}$$

where $n \geq 0$. It should be noted that once qualitative values [+] and [-] appeared simultaneously on the right hand side of this equation, $E(t_n)$ will always be an ambiguous value. For example, if $F(t_0) = [+]$ and $F(t_1) = [-]$, then the values of $E(t_1)$, $E(t_2)$, $E(t_3)$, ... will all be uncertain. Therefore, we do not adopt this equation in our qualitative representations.

According to the above discussion, the qualitative behaviour of I elements can be written as

$$E(t_2) = I \times (F(t_2) - F(t_1)). \tag{3-20}$$

Distribution Elements

The quantitative behaviour of a transformer has been characterised by Eqs. (3-6) and (3-7). In a real engineering system, the modulus b/a is always positive, so it can be neglected in qualitative representations. Thus, the qualitative description of TF elements is generalised as follows:

$$\begin{cases} E_{in} = E_{out} \\ F_{in} = F_{out} \end{cases}. \tag{3-21}$$

For the same reason, the qualitative representation of GY elements is:

$$\begin{cases} E_{in} = F_{out} \\ F_{in} = E_{out} \end{cases}. \tag{3-22}$$

Connections

Qualitative descriptions of 1-junctions and 0-junctions can also be directly obtained from their quantitative forms. For a 1-junction, it will be:

$$E_{in1} + E_{in2} + \ldots + E_{inm} = E_{out1} + E_{out2} + \ldots + E_{outn}, \tag{3-23}$$

$$F_{in1} = F_{in2} = \ldots = F_{inm} = F_{out1} = F_{out2} = \ldots = F_{outn}. \tag{3-24}$$

For a 0-junction, it will be:

$$E_{in1} = E_{in2} = \cdots = E_{inm} = E_{out1} = E_{out2} = \cdots = E_{outn}, \tag{3-25}$$

$$F_{in1} + F_{in2} + \cdots + F_{inm} = F_{out1} + F_{out2} + \cdots + F_{outn}. \tag{3-26}$$

There may be an argument as to why we do not neglect the parameters R, C and I in the qualitative representations because they are always positive. The reason is that the parameters are employed to represent the fault states for their corresponding elements in our fault diagnosis methodology (their usage is discussed in Chapter 6). Besides, it should be noted that all the primitives illustrated above are ideal. In real systems, a transformer may be coupled with the effects of resistance, capacitance and inertia. Hence, a real transformer will be represented by a set of bond graphs which combines the idealised elements TF, R, C, and I.

All the bond graph primitives used in this book are summarised in Table 3.3 with their symbols and general notations, which will be applied in all engineering domains.

Primitive	Symbol	Primitive	Symbol
Power Variable:		*Ideal Source:*	
Effort	E	Effort source	$S_e \longrightarrow$
Flow	F	Flow source	$S_f \longrightarrow$

Primitive	Symbol	Qualitative Representation	
Passive Element:			
Resistance	$\longrightarrow R$	$E(t) = R \times F(t)$	
Capacitance	$\longrightarrow C$	$F(t_2) = C \times (E(t_2) - E(t_1))$	
Inertial element	$\longrightarrow I$	$E(t_2) = I \times (F(t_2) - F(t_1))$	
Distribution:			
Transformer	$\longrightarrow TF \longrightarrow$	$E_{in}(t) = E_{out}(t), \quad F_{in}(t) = F_{out}(t)$	
Gyrator	$\longrightarrow GY \longrightarrow$	$E_{in}(t) = F_{out}(t), \quad F_{in}(t) = E_{out}(t)$	
Connection:			
Serial junction	$m: \diagdown\!\!\!1\!\!\!\diagup :n$	$E_{in1} + \cdots + E_{inm} = E_{out1} + \cdots + E_{outn}$ $F_{in1} = \cdots = F_{inm} = F_{out1} = \cdots = F_{outn}$	
Parallel junction	$m: \diagdown\!\!\!0\!\!\!\diagup :n$	$E_{in1} = \cdots = E_{inm} = E_{out1} = \cdots = E_{outn}$ $F_{in1} + \cdots + F_{inm} = F_{out1} + \cdots + F_{outn}$	

Table 3.3 Bond Graph Primitives and their Qualitative Representations

3.4.2   Model Building

An exact procedure for building bond graph models has been developed and has been widely used to help build physical models [Rosenberg, 1983; Biswas, 1992]. The procedure consists of seven steps.

1) Identify the dominant variables in the domain. In mechanics they are the flow variables; while in fluid, pneumatic, and electrical domains they are the effort variables.

2) Establish a junction for each instance of variables, for example, an 0-junction for each instance of dominant effort variables and a 1-junction for the flow ones.

3) Bond I elements to their 1-junction and C elements to their 0-junction.

4) Connect these junctions to each other using the complement junctions, TF and GY. Then, attach R elements whenever necessary.

5) Connect effort and flow sources to their proper junctions.

6) Find the grounding nodes which are zero-effort or zero-flow so that all bonds connected to them can be eliminated.

7) Simplify the graph by replacing any 2-port 0- or 1-junctions, which just pass power, with simple bonds. For example,

$$\longrightarrow 1 \longrightarrow \ = \ \longrightarrow, \qquad \text{and} \qquad \longrightarrow 0 \longrightarrow \ = \ \longrightarrow.$$

Fig. 3.2 shows an example of a separately excited DC motor with a resilient load, which is employed to explain the modelling procedure. R1 is the armature resistance of the motor, I1 is the armature inductance, I2 is the rotor inertia of the motor, C1 is a spring, I2 is the inertia of the load, and R2 is the friction of the load.

The modelling procedure begins with establishing the 0- and 1-junctions for the nodes a to e. Then, the energy storage elements, $I1$, $I2$, $C1$, and $I3$, are connected to these junctions, as shown in Fig. 3.3 (a). Next, these segments are connected by a GY element, and R elements. Then, the power source Se is inserted and the result is shown in Fig. 3.3 (b). Here, node c is a zero-effort junction, so the bonds connected to it are eliminated. In the end, the final bond graph Fig. 3.3 (c) is generated by simplifying Fig. 3.3 (b). In our approach, each bond has a number to mark its location.

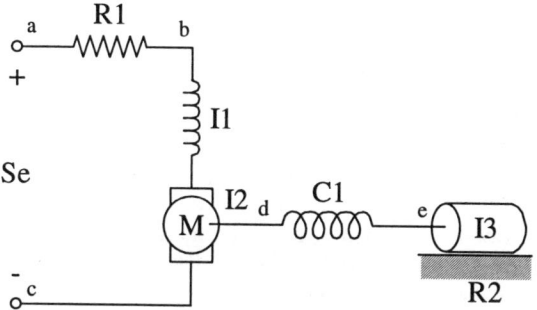

Fig. 3.2   Schematic Diagram of a Separately Excited DC Motor with a Resilient Load

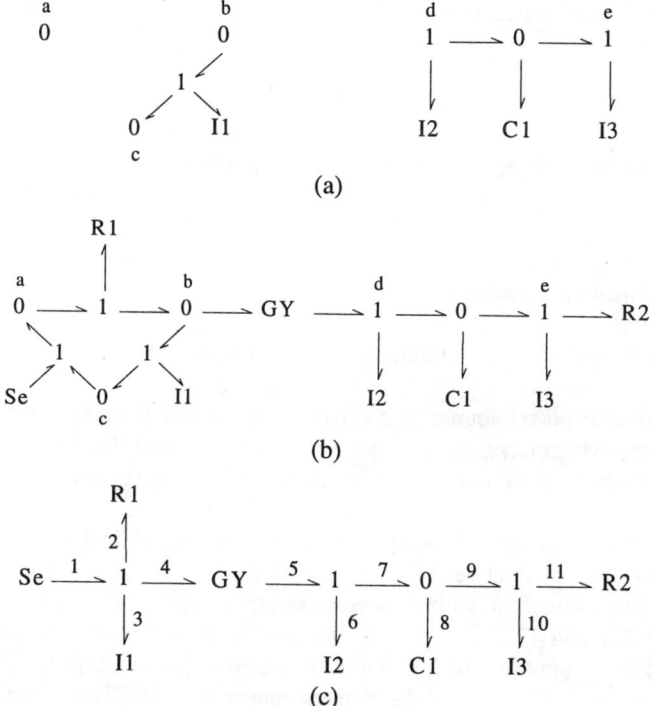

Fig. 3.3 Bond Graph Modelling Procedure for a DC Motor System

3.4.3   Qualitative Equation Generation

Typically, bond graph models are used to help generate differential equations, state-space equations, block diagrams, or signal-flow diagrams to systematise conventional modelling procedures. However, here a bond graph model is employed to produce a set of qualitative equations to represent important characteristics of a system. These equations connect constitutive element equations and contain all the necessary information (i.e. structural, behavioural, and functional information) about a physical system. In addition, qualitative equations relate components' behaviours to the behaviour of the whole system so that the relationships between components and their system can be analysed. As a result, a deep-level knowledge model can be represented by using the qualitative equations.

A schematic procedure is again developed here to guarantee the completeness for generating qualitative equations from a bond graph model. There are six steps.

1)   Find a power source in a bond graph model and the junction to which it is connected.

2)   Establish the constitutive equations for the junction.

3)   Establish the constitutive equations of the elements which are connected to this junction.

4)   Find next junctions which accept power from the previous ones and establish their constitutive equations.

5)   Repeat Steps 3) and 4) until no more junctions can be found.

6)   Find another power source, if it exists. Repeat Steps 1) to 5) and eliminate the equations which have already been stated. Then, repeat this step again until all elements in the bond graph are translated to qualitative equations.

Following this sequence, the bond graph model illustrated in Fig. 3.3 (c) can be transformed into qualitative equations. Firstly, the power source Se is found connected to a 1-junction, so the constitutive equations of the 1-junction are stated in Eqs. (3-27) and (3-28). Secondly, the constitutive formulas of the elements attached to the 1- junctions ($R1$, $I1$, and GY) are established in Eqs. (3-29) to (3-32). Thirdly, the next 1-junction and the element connected to it ($I2$) are found, and their qualitative representations are illustrated in Eqs. (3-33) to (3-35). Then, in the same way, the qualitative notations of a 0-junction and one more 1-junction and the elements connected to each of them are described in Eqs. (3-36) to (3-42). Finally, the results are shown as follows, where T is the sampling period.

$$
\begin{aligned}
E1(nT) &= E2(nT) + E3(nT) + E4(nT), & (3\text{-}27)\\
F1(nT) &= F2(nT) = F3(nT) = F4(nT), & (3\text{-}28)\\
E2(nT) &= R1 \times F2(nT), & (3\text{-}29)\\
E3(nT) &= I1 \times (F3(nT)\text{-}F3((n\text{-}1)T)), & (3\text{-}30)\\
E4(nT) &= F5(nT), & (3\text{-}31)\\
F4(nT) &= E5(nT), & (3\text{-}32)\\
E5(nT) &= E6(nT) + E7(nT), & (3\text{-}33)\\
F5(nT) &= F6(nT) = F7(nT), & (3\text{-}34)\\
E6(nT) &= I2 \times (F6(nT) - F6((n\text{-}1)T)), & (3\text{-}35)\\
E7(nT) &= E8(nT) = E9(nT), & (3\text{-}36)\\
F7(nT) &= F8(nT) + F9(nT), & (3\text{-}37)\\
F8(nT) &= C1 \times (E8(nT) - E8((n\text{-}1)T)), & (3\text{-}38)\\
E9(nT) &= E10(nT) + E11(nT), & (3\text{-}39)\\
F9(nT) &= F10(nT) = F11(nT), & (3\text{-}40)\\
E10(nT) &= I3 \times (F10(nT) - F10((n\text{-}1)T)), & (3\text{-}41)\\
E11(nT) &= R2 \times F11(nT). & (3\text{-}42)
\end{aligned}
$$

3.4.4   Deep-Level Knowledge Representation

So far a systematic method has been built to generate qualitative equations from the physical structure of a system. In this section, how the qualitative equations represent deep-level knowledge and some related problems will be discussed.

First of all, it can be found from Eqs. (3-27) to (3-42) that only variables $E1$ and $F1$ do not appear on the right hand side of the operator "=". That means, bond-1, which contains the variables $E1$ and $F1$, does not receive energy from any other junctions. In other words, energy is fed into the system from bond-1. Next, from Eqs. (3-27) and (3-28), it is noted that energy is distributed to bond-2, bond-3, and bond-4 via a common-flow junction. Then, Eqs. (3-29) to (3-32) show that a part of the energy is consumed by the element $R1$, some of the energy may be stored by the element $I1$, and some of it is delivered to other elements through a gyrator. Furthermore, the input effort $E1$ separates into the efforts $E2$, $E3$, and $E4$; where the $E2$ and $E3$ are decayed by the $R1$ and $I1$, while the $E4$ is translated to its corresponding flow $F5$. This describes the structural information about the system illustrated in Fig. 3.2. and shows that the qualitative equations indicate the elements' locations and their interactions.

Now, if we would increase the rotating speed of the load ($F11$) without decreasing the friction of the load ($R2$), then, from Eq. (3-42), the effort $E11$ must be increased. Then. from Eqs. (3-39), (3-36) and (3-33), the efforts $E9$, $E7$ and $E5$ should be increased. Again, from Eq.(3-32), the flow $F4$ should be increased. Next, from Eq. (3-28), the flow $F2$ should be increased, so, from Eq. (3-29), $E2$ should be increased as well. Finally, from Eq. (3-27), the input effort $E1$ should be increased. As a result, we know that if we want to increase the rotating speed of the load, then we must increase the input voltage. Here, the qualitative equations provide the conceptual function of the whole system in terms of the individual functions of its elements and junctions.

Also, let us assume the output torque of the motor ($E7$) is increasing, then, from Eq.(3-36), the effort $E8$ is increasing ($E8(nT) > E8((n-1)T)$). Then, from Eq. (3-38), the speed of the spring ($F8$) is positive, so, from Eq. (3-37), the rotating speed of the load ($F9$) is smaller than the rotating speed of the motor ($F7$). If we keep the torque of the motor constant, then, from Eq.(3-36), the spring effort $E8$ is constant ($E8(nT) = E8((n-1)T)$). Thus, from Eq. (3-38), the spring speed $F8$ is zero, while, from Eq. (3-37), the load speed $F9$ is equal to the motor speed $F7$. Then, following this inference route, it can be found that the load's speed will be bigger than the motor's speed when the torque of the motor is decreasing. Besides, from Eqs. (3-37) and (3-38), we also understand that the bigger value for $C1$ will cause a bigger speed drop between $F7$ and $F9$. This inference mechanism can be extended to the whole system to reason about system behaviours. Here, qualitative equations provide a formal representation for reasoning about the relationships between the elements' behaviours and the system behaviour.

From this description, it can be seen that the integration of bond graphs and qualitative equations provides an effective systematic method for constructing deep-level knowledge models. Although all the inferences mentioned above are conceived by human-thinking and described in natural language, this qualitative representation still exhibits high potential for building a generalised deep reasoning system.

For the purpose of formulating an automatic qualitative reasoning method, qualitative values and qualitative operators must be very well defined. Good definitions of qualitative values depend on not only the choice of their scale but, also the physical concepts represented by the values. A qualitative scale should divide a measurement space into several significant ranges to give adequate information for characterising the space. At the same time, it should be simple enough so that quantitative details can be ignored. On the other hand, the physical meaning of qualitative values is essential for conceptual inference. For instance, a qualitative value [+] may indicate a scalar quantity "positive", or denote a direction of velocity, or indicate the potential of force, etc. System parameters and some variables (i.e.

temperature) are not usually vectors, but some variables (i.e. velocity and force) also have direction. Accordingly, the physical meaning described by qualitative values must be defined together with the definition of their scale in order to avoid possible inference confusion.

Moreover, qualitative operators should also be defined to have reasonable operation results under the definitions of qualitative values. For example, when the qualitative value [+] indicates a "big" quantity and [++] indicates a "very big" quantity, then an operator + resulting in an answer: [+] + [+] + [+] + ... = [++] is reasonable. However, the same definition of operator + will lead to an unreasonable result when the value [+] denotes "normal increasing" for a variable and [++] denotes "rapidly increasing". In this situation, a more rational inference is: joining all normal states should produce a normal behaviour rather than an abnormal one. Thus, the operation should be [+] + [+] + [+] + ... = [+]. As explained, qualitative operators should be defined to support reasonable inferences relating to physical reality. Nevertheless, generality is also very important in the definitions of qualitative operators, otherwise reasoning with a complex system may require defining too many operators so that simplicity of qualitative reasoning is neutralised.

As discussed in Chapter 2, many attempts have been made to define powerful qualitative values and operators. The research in "order of magnitude" introduced by Raiman [1986] incorporated magnitude information with the relationship between variables, while the aspect of qualitative mathematics proposed by Struss [1988] also employed magnitude information to determine the continuations of qualitative behaviours. Fuzzy set theory and qualitative reasoning were combined by Shen and Leitch [1990] to increase the accuracy of qualitative models. The concept of causality was used by Iwasaki and Simon [1986] to reduce the complexity of qualitative inference operations. A representation method using qualitative vectors was developed by Morgan [1988] to resolve difficulties about integration and differentiation of qualitative values. In Chapter 6, definitions of qualitative values and operators together with their physical meaning will be used to construct an effective inference mechanism for the field of fault diagnosis.

An old problem, namely the completeness of system models, still exists in qualitative representation. Like most modelling methodologies, the detail-level (or accuracy) of a model must be decided by experience when modelling a complex system. In the above example, the effects of the back emf, the capacitance, and the axial friction of the motor were not considered. How these effects influence the completeness and the accuracy of the model can be explored by numerical simulations in conventional modelling methods. But, in qualitative reasoning, this problem appears more difficult because no numerical information can be used to judge detailed effects. In Chapter 4, the way in which the detail-level of a qualitative model influences the performance of a qualitative controller will be examined by experimentation.

3.4.5   Higher-Order Derivative

The higher-order derivative problem in qualitative reasoning was introduced by de Kleer and Bobrow [1984]. Suppose one drops a ball. At the moment the ball is released, it can't be moving, but immediately thereafter it is. At the moment of release it cannot have moved, has zero velocity, and negative acceleration. Qualitatively, $[x] = [0]$, $\partial x = 0$, and $\partial^2 x = -$. So $[x]$ becomes - even though $\partial x = 0$. The correct qualitative integration is $[x_{next}] = [x_{current}] + \partial^n x_{current}$ where $\partial^n x$ is the first non-zero derivative. The difficulty here is that the higher-order derivative $\partial^n x$ may not be known.

Fortunately, in the representation of qualitative equations, system order is easily determined from the variables mentioned in the equations. As in the DC motor case, the four difference equations (Eqs. (3-30), (3-35), (3-38), and (3-41)), which represent the behaviour of energy storage elements, indicate that the system is fourth order. Furthermore, each variable of this model is clearly related to others via the qualitative equations. If the initial conditions are given, the qualitative behaviour of the elements and the whole system is usually derivable. Thus, there is no need to devote too much effort to find system order and to infer the qualitative values of higher-order derivatives.

3.4.6   Simplification of Qualitative Equations

A successful modelling method in supervisory control systems requires using a general modelling procedure to build general models which can produce results with high-level knowledge reasoning and low-level detail computation. As discussed previously, the qualitative bond graph modelling method provides a systematic way to generate high-level knowledge models. In this section, a generic procedure will be developed to simplify the high-level qualitative representation for simple control needs.

Qualitative equations derived from system structure contain a lot of internal information about the system, where this information is very useful for high-level reasoning. However, internal information is not necessary for a simple feedback controller, since such a controller regulates a system only by referring to the system output and input commands. Therefore, qualitative equations can be simplified to describe directly the relationship between system inputs and outputs to generate control algorithms. Simplified qualitative equations will not infer the internal states of a system so that computational efficiency can be improved for performing on-line control. Also, in a qualitative environment, simplified equations can avoid many ambiguous problems during inference process.

A generic procedure for simplification of qualitative equations is developed as follows:

1) If system parameters are not obtained, then remove all system parameters R, C, and I from qualitative equations. If system parameters have already been obtained, then insert them into the qualitative equations.

2) Identify the variables of system inputs and outputs.

3) Replace the power variables with the smallest subscript variables according to the substitution relationships from the equations containing no "plus (+)" operators. It should be noted that the variables of system inputs and outputs cannot substitute for each other.

4) Cancel all the equations containing no "plus (+)" operators except those which include the input or output variables.

5) Replace all variables by the variables of system inputs and outputs according to the substitution relationships from all the equations.

6) Cancel the equations containing the variables which are not the ones of system inputs and outputs.

Here, the example illustrated in Fig. 3.2 is again employed to explain this procedure. Firstly, the system parameters are not considered, thus, R, C, and I are removed from the equations (3-27) to (3-42). Next, the input voltage $E1$ is identified as the system input and the rotating speed of the load $F10$ is chosen as the system output. Note that, from Eq. (3-40), the output variable $F10$ can be replaced by $F9$. Then, the power variables are replaced by the smallest subscript ones. Thus, the previous equations are now translated to the following:

$$
\begin{array}{llr}
E1(nT) & = & F1(nT) + F1(nT) - F1((n-1)T) + E4(nT), \qquad (3\text{-}43) \\
F1(nT) & = & F2(nT) = F3(nT) = F4(nT), \qquad (3\text{-}44) \\
E2(nT) & = & F1(nT), \qquad (3\text{-}45) \\
E3(nT) & = & F1(nT) - F1((n-1)T), \qquad (3\text{-}46) \\
E4(nT) & = & F5(nT), \qquad (3\text{-}47) \\
F1(nT) & = & E5(nT), \qquad (3\text{-}48) \\
F1(nT) & = & E4(nT) - E4((n-1)T) + E7(nT), \qquad (3\text{-}49) \\
E4(nT) & = & F6(nT) = F7(nT), \qquad (3\text{-}50)
\end{array}
$$

$$E6(nT) \quad = \quad E4(nT) - E4((n-1)T), \qquad\qquad (3\text{-}51)$$

$$E7(nT) \quad = \quad E8(nT) = E9(nT), \qquad\qquad (3\text{-}52)$$

$$E4(nT) \quad = \quad E7(nT) - E7((n-1)T) + F9(nT), \qquad (3\text{-}53)$$

$$F8(nT) \quad = \quad E7(nT) - E7((n-1)T), \qquad\qquad (3\text{-}54)$$

$$E7(nT) \quad = \quad F9(nT) - F9((n-1)T) + F9(nT), \qquad (3\text{-}55)$$

$$F9(nT) \quad = \quad F10(nT) = F11(nT), \qquad\qquad (3\text{-}56)$$

$$E10(nT) \quad = \quad F9(nT) - F9((n-1)T), \qquad\qquad (3\text{-}57)$$

$$E11(nT) \quad = \quad F9(nT). \qquad\qquad\qquad (3\text{-}58)$$

Then, all the equations containing no "plus (+)" operators are cancelled, and the equations are written as follows:

$$E1(nT) \quad = \quad F1(nT) + F1(nT) - F1((n-1)T) + E4(nT), \qquad (3\text{-}59)$$

$$F1(nT) \quad = \quad E4(nT) - E4((n-1)T) + E7(nT), \qquad\qquad (3\text{-}60)$$

$$E4(nT) \quad = \quad E7(nT) - E7((n-1)T) + F9(nT), \qquad\qquad (3\text{-}61)$$

$$E7(nT) \quad = \quad F9(nT) - F9((n-1)T) + F9(nT). \qquad\qquad (3\text{-}62)$$

Afterwards, all the variables are replaced by the system input or output variables according to the relationships represented in Eqs. (3-59) to (3-62), and then the equations containing the variables which are not the system input or output ones are cancelled. Finally, the simplified result is:

$$E1(nT) = 13F9(nT) - 22F9((n-1)T) + 16F9((n-2)T) - 6F9((n-3)T) + F9((n-4)T). \qquad (3\text{-}63)$$

In this example, qualitative representation shows a strong feature. It removes the difficulties and complexities relating to the units in measurement spaces, and makes the simplification sequence very easy and free. From Eqs. (3-43) to (3-58), it can be seen that there are many expressions equating different-unit variables such as $E2(nT)$ $= F1(nT)$, where $E2$ is voltage and $F1$ is current. Here, the qualitative equation actually equates their influence level to the system rather than their quantities. From this point of view, qualitative values should be described without units, otherwise qualitative equations cannot be recognised. Fortunately, this requirement of describing qualitative values is very close to the nature of human-thinking. In the human mind, the concept of a quantity is usually represented as "big", "small", "increasing", or "decreasing" which is not related to any units. For this reason, the interactions between system variables can be represented naturally via qualitative equations.

With respect to the simplified result, Eq. (3-63) shows directly the relationship between the system input and output. Obviously, this process is a fourth order system, so the system output is related to the system's input and history. Here, several integer number coefficients appear in the equation. These coefficients can be seen as the relative weight of each item rather than actual quantities. If the system parameters have been inserted into the qualitative equations during the simplification process, then the relationship among the variable weights in the simplified equation will be different from that described in Eq. (3-63) except all the parameters have an identical value. It can be inferred that the qualitative equations which involve system parameters is more accurate than the ones which contain no system parameters. Thus, an equation simplified with the consideration of system parameters can represent the relationship between the system input and output more accurately, and the control algorithm derived from this simplified equation can regulate the system more effectively. According to the simplified equation, the system output can be predicted easily from the system input and past data without referring to the internal states of the system. Chapter 4 will illustrate how to derive qualitative control algorithms from this type of expression, and compare the control performances of the control algorithms derived with and without considering system parameters.

In addition, if we consider the torque of the load ($E10$) as the system output, then it can be found from Eqs. (3-39) to (3-42) that the torque output is actually related to the speed of the load ($F9$). Higher speed causes more energy loss in the friction of the load, so keeping a constant torque output under various speed needs a differing power input. Consequently, both variables $E10$ and $F9$ are system outputs. Therefore, this instance becomes a single-input multi-output (SIMO) case. Now, following the simplification procedure Step 1 to Step 4, a set of qualitative equations can be obtained as follows:

$$
\begin{aligned}
E1(nT) &= F1(nT) + F1(nT) - F1((n\text{-}1)T) + E4(nT), & (3\text{-}64)\\
F1(nT) &= E4(nT) - E4((n\text{-}1)T) + E7(nT), & (3\text{-}65)\\
E4(nT) &= E7(nT) - E7((n\text{-}1)T) + F9(nT), & (3\text{-}66)\\
E7(nT) &= E10(nT) + F9(nT), & (3\text{-}67)\\
E10(nT) &= F9(nT) - F9((n\text{-}1)T). & (3\text{-}68)
\end{aligned}
$$

Then, replacing all the variables by the input and output ones, according to the relations described in these equations, leads to the final result:

$$E1(nT) = 11E10(nT) - 11E10((n\text{-}1)T) + 5E10((n\text{-}2)T) - E10((n\text{-}3)T) + 2F9(nT). \quad (3\text{-}69)$$

From this equation, the output torque can be predicted by referring to the input voltage $E1$, the current speed $F9$, and some past system behaviour. From the result that a simplified qualitative equation successfully describes a SIMO system, we can infer that a set of such equations can also be used to represent a multi-input multi-output (MIMO) system.

Let us consider another example with taking account of system parameters. Fig. 3.4 shows a fluid system and its bond graph model. The qualitative equations which represent this system are shown in Eqs. (3-70) to (3-87).

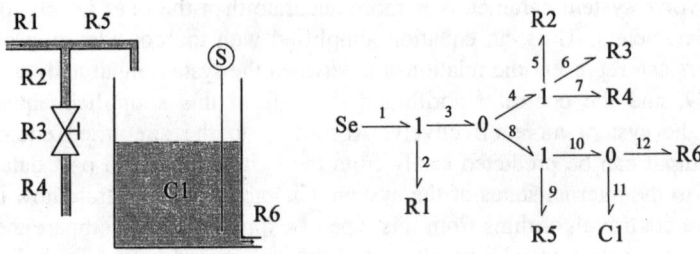

Fig. 3.4 A Fluid System and its Bond Graph Model

$$E1(nT) = E2(nT) + E3(nT), \tag{3-70}$$
$$F1(nT) = F2(nT) = F3(nT), \tag{3-71}$$
$$E2(nT) = R1 \times F2(nT), \tag{3-72}$$
$$E3(nT) = E4(nT) = E8(nT), \tag{3-73}$$
$$F3(nT) = F4(nT) + F8(nT), \tag{3-74}$$
$$E4(nT) = E5(nT) + E6(nT) + E7(nT), \tag{3-75}$$
$$F4(nT) = F5(nT) = F6(nT) = F7(nT), \tag{3-76}$$
$$E5(nT) = R2 \times F5(nT), \tag{3-77}$$
$$E6(nT) = R3 \times F6(nT), \tag{3-78}$$
$$E7(nT) = R4 \times F7(nT), \tag{3-79}$$
$$E8(nT) = E9(nT) + E10(nT), \tag{3-80}$$
$$F8(nT) = F9(nT) = F10(nT), \tag{3-81}$$
$$E9(nT) = R5 \times F9(nT), \tag{3-82}$$
$$E10(nT) = 0, \tag{3-83}$$
$$E11(nT) = E12(nT), \tag{3-84}$$
$$F10(nT) = F11(nT) + F12(nT), \tag{3-85}$$
$$F11(nT) = C1 \times (E11(nT) - E11((n-1)T)), \tag{3-86}$$
$$E12(nT) = R6 \times F12(nT). \tag{3-87}$$

In this example, concerned is the relationship between the input pressure (*E1*) and the liquid level in the tank (*E11*). Thus, *E1* can be seen as the input variable and *E11* the output variable. Further, the parameter *R3* is an interesting factor in the interrelations of *E1* and *E11*, so it should be kept for the simplified equation. Other parameters in this system are constant and not so substantial as *R3*. Therefore, they are ignored in this example. Through the Step 3) of the simplification process, the previous equations are translated to the following:

$$
\begin{aligned}
E1(nT) &= F1(nT) + E3(nT), & (3\text{-}88) \\
F1(nT) &= F2(nT) = F3(nT), & (3\text{-}89) \\
E2(nT) &= F1(nT), & (3\text{-}90) \\
E3(nT) &= E4(nT) = E8(nT), & (3\text{-}91) \\
F1(nT) &= F4(nT) + F8(nT), & (3\text{-}91) \\
E3(nT) &= F4(nT) + R3 \times F4(nT) + F4(nT), & (3\text{-}93) \\
F4(nT) &= F5(nT) = F6(nT) = F7(nT), & (3\text{-}94) \\
E5(nT) &= F4(nT), & (3\text{-}95) \\
E6(nT) &= R3 \times F4(nT), & (3\text{-}96) \\
E7(nT) &= F4(nT), & (3\text{-}97) \\
E3(nT) &= F8(nT) + 0, & (3\text{-}98) \\
F8(nT) &= F9(nT) = F10(nT), & (3\text{-}99) \\
E9(nT) &= F8(nT), & (3\text{-}100) \\
E10(nT) &= 0, & (3\text{-}101) \\
E11(nT) &= E12(nT), & (3\text{-}102) \\
F8(nT) &= F11(nT) + E11(nT), & (3\text{-}103) \\
F11(nT) &= E11(nT) - E11((n\text{-}1)T), & (3\text{-}104) \\
E11(nT) &= F12(nT). & (3\text{-}105)
\end{aligned}
$$

Then, cancel the equations containing no "+" operators and I/O variables. The previous equations can be simplified to the following:

$$
\begin{aligned}
E1(nT) &= F1(nT) + E3(nT), & (3\text{-}106) \\
F1(nT) &= F4(nT) + F8(nT), & (3\text{-}107) \\
E3(nT) &= F4(nT) + R3 \times F4(nT) + F4(nT), & (3\text{-}108) \\
E3(nT) &= F8(nT) + 0, & (3\text{-}109) \\
F8(nT) &= F11(nT) + E11(nT), & (3\text{-}110) \\
F11(nT) &= E11(nT) - E11((n\text{-}1)T), & (3\text{-}111)
\end{aligned}
$$

Then, through the Steps 5) and 6) of the simplification process, the final result can be obtained as:

$$
E1(nT) = \frac{5+2R3}{2+R3} \times (2E11(nT) - E11((n\text{-}1)T)). \qquad (3\text{-}112)
$$

This simplified equation represents clearly the relations between $E1$, $E11$ and $R3$. It shows that the liquid level of the tank can also be regulated by tuning the parameter $R3$.

3.5   DISCUSSION

In this chapter, two different approaches to using qualitative representation for problem solving, i.e. fuzzy set theory and qualitative reasoning, have been compared. Their main differences are classified in terms of the fundamental motivation of development and the formulation of knowledge representation. Generally speaking, the fuzzy set theory provides a subjective form which represents system knowledge with a rule base to describe the relationships between a system input and output in natural language, where the rule base is acquired from domain expert experience or self-learning. However, qualitative reasoning is a paradigm which represents system knowledge based on the governing physical laws of the system, where the system model is built through analysing system structure. According to the comparison, qualitative reasoning seems more suitable for describing system structure information for high-level cause-effect reasoning in supervisory control systems.

For the purpose of developing a generic modelling method to systematically construct qualitative models, bond graph modelling language is integrated with qualitative reasoning. Bond graph methodology is a formal modelling scheme used to build models for all engineering systems from limited primitives. In this chapter, a number of qualitatively constitutive equations of bond graph primitives were defined to combine bond graphs with qualitative reasoning. A general procedure for producing bond graph models from system structure was introduced, while a further general procedure was developed to derive a set of qualitative equations from a bond graph model. These qualitative equations connect element constitutive equations according to their interconnections and offer an explicit formulation to represent the knowledge about system structure. With the qualitative representation, system function, behaviour and their interactions can be characterised via qualitative operations. The use of this compositional modelling methodology successfully matches system structures to their functions and provides three further advantages:

- The behaviour of cross-field engineering systems can be analysed in a unified language without need to use numerical details.

- The deep-level knowledge about a system can be represented in a formal mathematical form.

The generalised modelling procedure and knowledge representation form make it possible to automatically generate deep models.

This chapter also developed a generic procedure for simplification of qualitative equations to describe directly the relationships between a system input and output. The simplified equations allow one to predict system behaviour by reasoning only about the system input and history, rather than inferring the internal states of the system. Therefore, some ambiguous problems caused by qualitative operations and the difficulties of assigning causality inside system models can be avoided so that qualitative reasoning can be applied more widely to tasks where the internal states are unnecessary. This representation also shows a noteworthy feature. That is, through the combination of the simplification procedure with bond graph modelling and qualitative equation generation procedures, control algorithms for qualitative controllers can be generated automatically from a system structure without using numerical details about the system. Although the simplification method has successfully simplified the examples illustrated in this and the following chapters, a formal proof of its effectiveness is still needed.

The methodologies of qualitative bond graph modelling described in this chapter provide a basis for qualitative model-based reasoning. The aspects of qualitative supervisory control systems, such as qualitative control, self-tuning, and fault diagnosis, can be thus performed according to the qualitative representations illustrated in this chapter. The next chapter will discuss how to combine qualitative models with quantitative data to produce a high performance qualitative controller. Chapter 5 will introduce a self-tuning method to further improve the control performance. Chapter 6 will propose a fault diagnosis method which can locate system faults via structural information contained in qualitative equations.

CHAPTER 4

HYBRID QUALITATIVE AND QUANTITATIVE CONTROL

4.1 INTRODUCTION

For many years, most of the growing interest in qualitative reasoning has been concerned with the analyses of physical systems, such as qualitative simulation and fault diagnosis [Weld and de Kleer, 1990]. However, applying qualitative knowledge to primitive control tasks has received less attention. Two considerable advantages show that a qualitative control method is worthy to be developed. On the one hand, qualitative control method can easily relate high-level knowledge to low-level control algorithms, so it can regulate a system based on the understanding of the system structures. The control action of such a control method is much closer in spirit to the behaviour of human operators than that of conventional control methods. Accordingly, it provides higher robustness and safety for industry control. On the other hand, as discussed in the previous chapter, a qualitative control algorithm is derivable from system structure without numerical details via the formal procedure of the qualitative bond graph modelling scheme. Thus, an automatic design tool for feedback control systems can be built without using deep mathematical techniques.

The first attempt at the idea of qualitative control was proposed by Francis and Leitch [1984], where a real-time shell for intelligent feedback control called ARTIFACT was developed in PROLOG and C languages. Knowledge is represented by a set of conditional goal/subgoal pairs in the form [[Antecedent], [Subgoal, Goal]]. These pairs are used to represent subsystem relationships within a complex system, and taken together according to a relational model. System states and control actions in these PROLOG clauses are described in linguistic terms such as "too_low", "correct", "increasing", and "steady". Then, an inference mechanism works from the current system states by finding a set of clauses whose consequents lead the system to the desired goal states. This method was illustrated on a 2-coupled tanks liquid level control rig, with extension to an *n*-coupled tanks case. This approach opened a new knowledge-based control technique compatible with human-like cognition and reasoning.

The phrase "Qualitative Control" first appeared in the paper of Clocksin and Morgon [1986]. A number of qualitative controllers based on different control strategies (i.e. PID control, target control, and delta-modulation control ...etc.) were derived and demonstrated on the 2-coupled tanks liquid level control rig using Lisp programming language. Some significant variables of system input and output, such as liquid level of Tank 2, change rate of the level in Tank 2, and pumping rate, are used to represent the semantics of the system. The use of variables is determined by users according to the individual needs of the control strategies. These variables are related by the notions of each control strategy to produce various control rules. System states and control actions are described by qualitative values +, 0, and -, while control rules are represented in the form of qualitative equations. Current system states are inserted into the control rules, and the corresponding controller output is then derived via qualitative operations. As discussed, in this approach, the control strategies and their control rules are assigned by designers, therefore the high-level inference about system behaviours is implemented by designers rather than the controller itself. From this point of view, this qualitative control methodology is quite similar to the methods of conventional control.

Another interesting approach to qualitative control was proposed by Far [1989], where a smart robot governed by a hierarchical qualitative controller was used as a human operator to operate the inlet and outlet valves to regulate the liquid level for a liquid level control rig. Knowledge about the system is represented by qualitative signal flow graphs, while system states and control actions are described in linguistic terms, i.e. "slower", "faster", and "open", etc. An inference mechanism deduces the control actions by using a similar method to that proposed by Francis and Leitch. The derived control commands guide the robot to open or close the inlet and outlet valves to control the liquid level. In contrast to the former approaches, this qualitative control methodology was employed at a higher hierarchical level rather than a lower control level.

These approaches have demonstrated several fundamental ideas about qualitative control. In this chapter, attempts are made to enhance the practicability of qualitative control. Firstly, a new representation of a control algorithm will be developed. With this representation, the controller can directly acquire system knowledge from qualitative bond graph models. Moreover, this representation is in contrast to the use of heuristic rules. Mathematical computations can be used in this representation so that current procedural computer languages can more easily deal with control algorithms to derive the controller output. Secondly, quantitative information about the system input and output will be associated with qualitative control algorithms in real-time to offset two weaknesses of qualitative control — one is that ambiguities usually occur in qualitative reasoning processes, the other is that some accurate information is lost in converting system quantities to simple qualitative values. As a result, control performance can be improved by applying this hybrid qualitative and quantitative control method. The application of this method will be illustrated by SISO and MIMO coupled tanks liquid level control rigs, operating in real-time using Modula-2 software code.

4.2   CONFIGURATION OF THE CONTROLLER

The block diagram illustrated in Fig. 4.1 shows the basic configuration of the hybrid qualitative and quantitative control method, including its knowledge acquisition process.

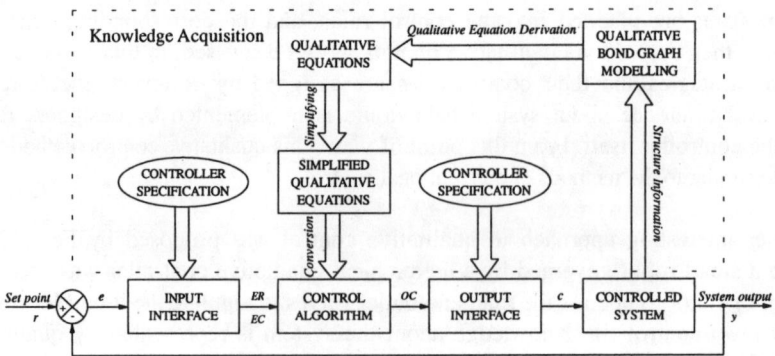

Fig. 4.1 Configuration of Hybrid Qualitative and Quantitative Control Method

As discussed in Chapter 3, the controlled system is modelled by a qualitative bond graph modelling method in terms of system structure. Then, the bond graph model is translated into a set of qualitative equations via a systematic procedure. Thus, designers should decide the input and output variables of the system. According to this decision, the qualitative equations are simplified via a generic simplification procedure to directly describe the relationships between system input and output. In this simplification process, all the state variables which haven't been chosen as the system input or output ones are omitted, since they are not necessary for the control algorithm. Furthermore, in this control method, input variables of the control algorithm are system errors and error changes, while the output variables are controller output changes. Consequently, state variables in the simplified equation(s) should be transferred to the variables of system error, error change, and controller output change. Thus, the qualitative control algorithm is obtained. This description shows that the control algorithm can be systematically derived from system structure without using any numerical details about the system, and therefore, a qualitative controller can be designed once the system structure is specified.

The only numerical data needed in this controller design process is the specifications of A/D and D/A converters. For example, a 12 bit converter can divide a continuous space into maximum counts of 4096 (2^{12}). The value description of the divisions could be 0 ~ +4095 or -2048 ~ +2047. This information is necessary for a digital controller to correctly acquire and send data.

The input interface calculates system errors and error changes according to the measurement data read from A/D converter, and then maps them into a qualitative space. The control algorithm derives controller output changes from these data. Then, the output interface establishes control actions by referring to the controller output changes. Finally, the D/A converter converts this digital data into analogue signals to regulate the system. Here, the roles of input and output interfaces are very similar to those of fuzzificaion and defuzzication interfaces in fuzzy logic controllers.

4.3   THE MERGING OF QUANTITATIVE INFORMATION

Most qualitative reasoning approaches, including previously proposed qualitative control methods, employ a coarse measurement scale for quantity representation to reduce the complexity of the reasoning systems. But, at the same time, this representation also reduces the precision of the statements about systems' situations. Unfortunately, especially for a controller, precision is one of the most important criteria. Poor control performance will be caused by the lack of precise quantitative information. One of the methods which can be used to improve the performance of qualitative controllers is to represent system situations with a finer partitioned measurement subspace. A more finely subdivided measurement scale provides more quantitative information so that more small variations of a system can be distinguished. Therefore, qualitative controllers can more accurately adjust control commands to regulate the systems.

What is the finest possible set of measurement values? In a digital control system, the resolution of its A/D D/A converter is the limit. As discussed previously, a 12 bit converter can divide a measurement space into a maximum of 4096 partitions. A measurement scale with so many subdivisions can provide a large amount of quantitative information. Yet, the problem is how to explicitly describe these partitions. In qualitative reasoning, partitions are usually represented by the terms drawn from an extremely limited set of symbols (i.e. +, 0, -, etc.) or linguistic descriptors (i.e. high, correct, low, etc.). Thus, expressing a large number of partitions by these limited symbols or linguistic descriptors and defining corresponding operators is very difficult. To resolve this problem, real numbers are employed to help describe these partitions, while the common operators for real numbers are used for their operations. As a result, this data representation is closer to quantitative rather than qualitative. However, will applying the quantitative measurement scale incur the same difficulties which occur in conventional quantitative approaches? In practice, these difficulties can be avoided if the quantitative scale is only used to measure system inputs and outputs. That is because most difficulties in conventional quantitative approaches arise from the need to evaluate the system parameters correctly. So, in the modelling stage, qualitative modelling methods can be employed to avoid the needs of system parameters, while, at the on-line control stage, quantitative measurement can be interpolated to improve control performance. Thus, the strengths of quantitative and qualitative approaches can be integrated to compensate each other's weaknesses in order to achieve an effective control methodology.

In addition, the qualitative control algorithms used in this hybrid control method are expressed in the form of mathematical equations. In contrast to representation by heuristic rules, making use of a finer measurement value set will not increase the complexity of the control algorithms. Therefore, quantitative data can be associated easily with the control algorithms described by the qualitative equations. Consequently, applying the qualitative equations representation makes the hybrid qualitative and quantitative control method practicable.

4.4   ALGORITHMS OF THE CONTROLLER

4.4.1   Input Interface

The main function of the input interface is to evaluate system errors and error changes from system outputs, and then map them into a corresponding universe of discourse for the qualitative control algorithm. Here, system error and error change are defined as

$$\hat{e}(nT) = set \text{ point}(nT) - \text{system output}(nT), \tag{4-1}$$

$$\hat{e}c(nT) = \hat{e}(nT) - \hat{e}((n-1)T), \tag{4-2}$$

where T is the sampling period, $\hat{e}$ is the system error, and $\hat{e}c$ denotes error change. All values contained in these equations have been obtained by an A/D converter.

As discussed in Chapter 3, control algorithms relate system output variables to input ones by their influence-level on the system. Consequently, error and error change should be mapped firstly into their "influence-level" on the system, and then they can be inserted into control algorithms to derive control actions. Here, system error is mapped via the following rule:

$$ER(nT) = \begin{cases} \dfrac{\hat{e}(nT)}{SF_e \times \text{Max. positive } \hat{e}(nT)} & \text{, when } \hat{e}(nT) \geq 0, \\[4mm] \dfrac{-\hat{e}(nT)}{SF_e \times \text{Max. negative } \hat{e}(nT)} & \text{, when } \hat{e}(nT) < 0, \end{cases} \qquad (4\text{-}3)$$

where ER is the mapped system error and SF_e is the error scaling factor. Here, the values of maximum positive and negative errors can be obtained by referring to the set-point and its measurement range. For example, an A/D converter could specify a measurement range 0 ~ +4096 and the set-point to be given by 1800. Thus, the maximum positive error will be 1800 - 0 = 1800, while the maximum negative error will be 1800 - 4096 = -2296. Dividing a system error by the maximum value of error transfers the error to a relative quantity which can be seen as its "influence-level" on the system. Likewise, error changes can be mapped to their relative values by the following notation:

$$EC(nT) = \begin{cases} \dfrac{\hat{e}c(nT)}{SF_{ec} \times \text{Max. positive } \hat{e}c(nT)} & \text{, when } \hat{e}c(nT) \geq 0, \\[4mm] \dfrac{-\hat{e}c(nT)}{SF_{ec} \times \text{Max. negative } \hat{e}c(nT)} & \text{, when } \hat{e}c(nT) < 0, \end{cases} \qquad (4\text{-}4)$$

where EC is the mapped error change and SF_{ec} is the error change scaling factor. Unlike the maximum error, the value of maximum error change is very difficult to estimate in advance with qualitative information. Accordingly, its value is measured on-line and continuously updated during the whole control process.

In the input interface, the scaling factors SF_e and SF_{ec} have effects on controller sensitivities to system error and error change. They also play a critical role in influencing control performance, such as rise time, settling time, and stability, etc. How to select the best values for scaling factors in a qualitative environment is one of the main problems for constructing an effective controller. In the next chapter, an automatic tuning method will be developed to adjust the scaling factors on-line and their effects on a controller will be discussed further.

4.4.2   Control Algorithm

The control algorithms for this hybrid control method are derived by simplifying the qualitative equations used to represent the qualitative models of physical systems. In Chapter 3, simplification of qualitative equations has been discussed and it can be seen that the simplified qualitative equations directly represent the relationships between system input and output variables. However, the control algorithms used here are required to describe the relationships between system errors, error changes, and system input changes. Therefore, simplified qualitative equations should be transferred to express interactions among errors, error changes, and output changes. Here, a simple example is given to help explain the derivation of the control algorithms.

Derivation of Control Algorithms

Fig. 4.2 illustrates a liquid level tank and its bond graph model. Its qualitative equation representation is stated in Eqs. (4-5) to (4-11). In this case, the inlet liquid pressure $E1$ is used to control the liquid level in Tank C.

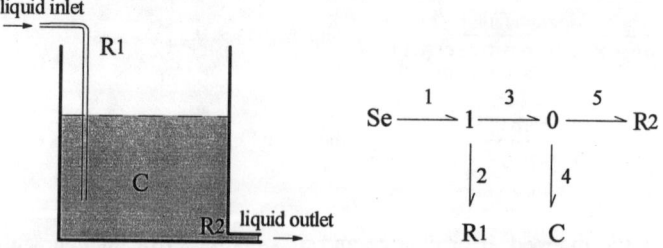

Fig. 4.2 Liquid Tank and its Bond Graph Model

$$E1(nT) = E2(nT) + E3(nT), \tag{4-5}$$
$$F1(nT) = F2(nT) = F3(nT), \tag{4-6}$$
$$E2(nT) = R1 \times F2(nT), \tag{4-7}$$
$$E3(nT) = E4(nT) = E5(nT), \tag{4-8}$$
$$F3(nT) = F4(nT) + F5(nT), \tag{4-9}$$

$$F4(nT) = C \times (E4(nT) - E4((n-1)T)), \qquad (4\text{-}10)$$
$$E5(nT) = R2 \times F5(nT). \qquad (4\text{-}11)$$

Using the simplification procedure explained in Chapter 3, the system parameters are first cancelled. Then, the system output variable is identified as $E4$, while the input variable is $E1$. Next, all the state variables in these qualitative equations are considered representing the quantities of error. In this situation, $E4$ describes the error in the liquid level of Tank C, while $E1$ can be seen as the input pressure error compared to the standard pressure which corresponds to the set-point of the liquid level. According to the basic concept about control tasks, the input pressure should tend towards its standard value so that the liquid level can be kept at the set-point. The input pressure error $E1$ can be regarded as the controller output change. This inference process is stated as follows. Firstly, the relations between system input and controller output are

> *standard input pressure   =   standard controller output,*
> *current input pressure   =   current controller output.*

Secondly, owing to the definition of system error, the input error can be written as

> *input pressure error = standard input pressure - current input pressure.*

According to the previous discussion, the controller output changes should let the controller output tend towards its standard value as follows:

> *standard controller output*
> *= current controller output + controller output change.*

As a result, the controller output change is

> *controller output change*
> *= standard controller output - current controller output*
> *= standard input pressure - current input pressure*
> *= input pressure error.*

From the above inference, the variables of system error and controller output change are identified as $E4$ and $E1$. The next step is to identify the error change of the system. From the definition of error change, Eq. (4-10) clearly indicates that $F4$ is the error change. Besides, from Eq. (4-9), the system error $E4$ can be replaced by the variable $E3$. Thus, the qualitative equations are simplified

$$E1(nT) = 2E3(nT) + F4(nT). \qquad (4\text{-}12)$$

Moreover, the symbols *E1*, *E3*, and *F4* can be replaced by that of *OC* (controller output change), *ER* (system error), and *EC* (error change) to more clearly represent the control algorithm

$$OC(nT) = 2ER(nT) + EC(nT). \tag{4-13}$$

Eq. (4-13) demonstrates a very clear control strategy, where controller output change can be inferred according to the system error and error change. Through this process, the control algorithm for the rig is obtained systematically.

Derivation of Controller Output Change

The basic idea about evaluating a controller output change is to derive the controller output which will approximate both the system error and error change to zero. Based on this idea, the inference mechanism of the control algorithm firstly refers to the current system error to deduce the next possible error change which will approximate the system error to zero. According to the deduced error change, the next possible system error can be inferred. The next controller output change can be derived through inserting the predicted error and error change into the control algorithm. This detailed inference procedure is illustrated in Fig. 4.3.

Fig. 4.3 shows a typical unit step response of a feedback control system. It is assumed that the system situation is at State 1, where the system error is bigger than the absolute value of the maximum negative error change. The inference mechanism will thus let the system approach the set-point at the maximum possible changing rate of maximum negative error change. Ideally, this system will then achieve State 2 at the next sampling period with a system error of $a + b$. Thus, inserting the error $a + b$ and maximum negative error change into the control algorithm will result in the controller output change for the situation of State 1.

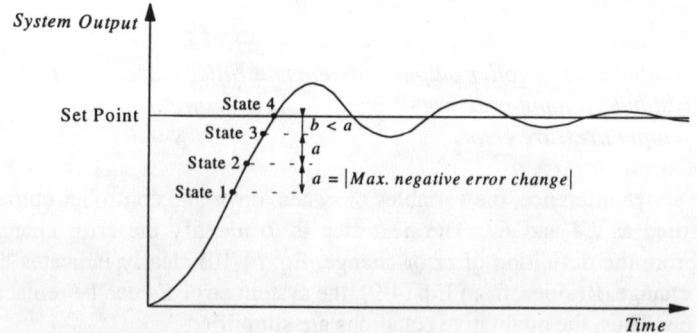

Fig. 4.3 Typical Unit Step Response of a Feedback Control System

At State 2, the system error is still bigger than the absolute value of maximum negative error change, so the system is kept approaching the set-point at the same maximum rate which will bring the system to State 3. This time, the error b and the maximum negative error change will be inserted into the control algorithm to explore the controller output change. At State 3, the system error is smaller than the absolute value of maximum negative error change, so keeping the same approaching rate will cause overshoot in the system at the next sampling period. Consequently, the following error change rate should be reduced. In an ideal case, setting the error change rate to $-b$ can lead the system to the set-point where the system error is zero. The next system error 0 and error change $-b$ can be inserted into the control algorithm to derive the controller output change. This inference process can also be applied when the error is negative. Theoretically, the system will finally achieve the steady-state at the set-point.

In conclusion, the prediction of next system error and error change can be summarised as follows

$$\hat{e}c(nT) = \begin{cases} \text{Max. negative } \hat{e}c(nT)\,, \text{ when } -\hat{e}((n-1)T) \le \text{Max. negative } \hat{e}c(nT), \\[2mm] \text{Max. positive } \hat{e}c(nT)\,, \text{ when } -\hat{e}(n-1)T) \ge \text{Max. positive } \hat{e}c(nT), \\[2mm] -\hat{e}((n-1)T) \qquad\quad \text{, when other states,} \end{cases} \tag{4-14}$$

$$\hat{e}(nT) = \hat{e}((n-1)T) + \hat{e}c(nT). \tag{4-15}$$

The predicted values of $\hat{e}c(nT)$ and $\hat{e}(nT)$ will be mapped into their corresponding values of $EC(nT)$ and $ER(nT)$ via Eqs. (4-3) and (4-4). They can be inserted into the control algorithm, such as the one described in Eq. (4-13), to infer the controller output change $OC(nT)$. The derived value of $OC(nT)$ will be fed into the output interface to produce controller output. This concept can also be applied to a ramp input or other style inputs. Here, the reference input of the system will be a variable value rather than a fixed one. The next error and error change are predicted according to the variable reference input rather than the fixed step input.

4.4.3   Output Interface

The function of the output interface is to convert the relative controller output change OC generated by the control algorithm into a practical controller output. The conversion equation is expressed as follows:

$$CO(nT) = \begin{cases} \dfrac{OC(nT)}{SF_O \times Max.\ OC} \times \text{positive output range} + A((n-1)T), \text{ when } OC(nT) \geq 0, \\[4mm] \dfrac{OC(nT)}{SF_O \times Max.\ OC} \times \text{negative output range} + A((n-1)T), \text{ when } OC(nT) < 0, \end{cases} \qquad (4\text{-}16)$$

where CO is the controller output, SF_O is the output scaling factor, and A denotes the average value of previous controller outputs. Here, the maximum possible controller output change (*Max. OC*) can be obtained through inserting maximum error and error change into the control algorithm. The output ranges are determined by both the value of A and the measurement range of D/A converter. For example, if a D/A converter provides a measurement range of $0 \sim +4096$, then the positive output range will be $4096 - A((n-1)T)$ while the negative output range will be $A((n-1)T) - 0 = A((n-1)T)$. This equation provides a scaling conversion which will map the relative controller output change into a corresponding universe of discourse of controller output (i.e. $0 \sim +4096$). The output scaling factor SF_O has an effect on the gain of the controller. It also plays a critical role in influencing the stability of the overall feedback control system. The detailed effects and an automatic tuning method for SF_O will be discussed in Chapter 5.

The purpose of adding the previous controller outputs' average A to the controller output change is to avoid steady-state error. In most control cases, controller output change is added only to the prior controller output rather than to the average of all previous ones. The reason for making use of the average value is that it will reduce the activity of the controller output. Excessive activity in the controller output will shorten the life time of actuators and that is not desirable in a practical system. However, making use of the arithmetic average value will also reduce the controllers' response speed so that the controller cannot track rapid changes very well. In order to overcome this problem, the technique of a recursive moving exponential window [Young, 1984] is employed to obtain a better average value of past controller outputs.

A moving exponential window provides an exponential decay function which will assign weights of past controller outputs to the value of A, and the weights will decay at an exponential rate. The recursive equations of the moving exponential window are written in the form

$$A(nT) = A((n-1)T) + P(nT) \times [CO(nT) - A((n-1)T)], \qquad (4\text{-}17)$$

$$P(nT) = \frac{1}{\alpha(nT)} \times \left(P((n-1)T) - \frac{P((n-1)T)^2}{\alpha(nT) + P((n-1)T)} \right),$$

$$\alpha(nT) = \lambda \times \alpha((n-1)T) + (1-\lambda) \times \delta,$$

with typical values of $\alpha(0) = 0.95$ and $\lambda = 0.99$. Here, δ is a forgetting factor whose value is between 0 and 1. It will make the output interface gradually "forget" former controller outputs but progressively utilise later ones. Therefore, the controller developed here will adapt more effectively to rapid changes of a system. A smaller δ will let the effects of past controller outputs decay faster, and vice versa.

In addition, the initial value of A is an important factor in the settling time of a control system. A well-settled initial value $A(0)$ will reduce the settling time for a system. Accordingly, a simple self-learning method is developed to estimate an adequate value of $A(0)$. Firstly, the output interface stores two different set-points and their corresponding steady-state controller output values, which are learned from operation experience. The steady-state controller output can be regarded as an adequate value of $A(0)$ corresponding to its set-point. Then, the initial values of A of various set-points are estimated via linear interpolation or extrapolation on the stored values. However, the relationship between set-points and steady-state controller outputs could be nonlinear. At this stage, the output interface employs the technique of curve fitting to approximate the nonlinear relationship with more learning data to derive more accurate initial values of A.

Nevertheless, adding the average value of the controller output to controller output changes will introduce the phenomenon of integrator windup which will cause actuators to saturate and a large overshoot to the control system. Therefore, the technique of anti-windup is necessary to avoid this problem. The anti-windup scheme used here is that the controller output changes are added to the value of $A(0)$ under the conditions of $\hat{e}c((n-1)T) < 0$, $0 < \hat{e}((n-1)T)$, and |*Max. negative* $\hat{e}c((n-1)T)| < \hat{e}((n-1)T)$. The function of the moving exponential window will keep on working without being interrupted by the action of anti-windup during the whole control process.

4.4.4   An Overview on the Algorithms

The overall data scaling mapping process of the hybrid control method is illustrated in Fig. 4.4 which is used to help interpret all the algorithms of the controller.

Let us assume that a system is in the situation of $y((n-1)T)$. This system output is measured by a sensor and than converted into the digital measurement space by a transducer and an A/D converter. Now, the system output $y((n-1)T)$ will be transferred into the representation of system error. Then, the current system error is compared with the past system error to obtain the error change. After this, the system error and the error change are mapped into their corresponding space as *ER((n-1)T)* and *EC((n-1)T)* through the scaling mappings described in Eqs. (4-3) and (4-4). Then, the mapped data *ER((n-1)T)* and *EC((n-1)T)* are inserted into the control algorithm to represent the current system state.

Meanwhile, the inference mechanism of the controller judges that, in this situation, the next error change should be positive so that the system error can be forced to approach zero. When the absolute value of the current system error is bigger than the maximum positive error change, then the next error change $\hat{e}c\,(nT)$ should be the maximum positive error change. Thus, the next system error $\hat{e}\,(nT)$ can be predicted by Eq. (4-15). Then, they can be mapped to their corresponding values *EC(nT)* and *ER(nT)* which will be inserted into the control algorithm to produce the controller output change.

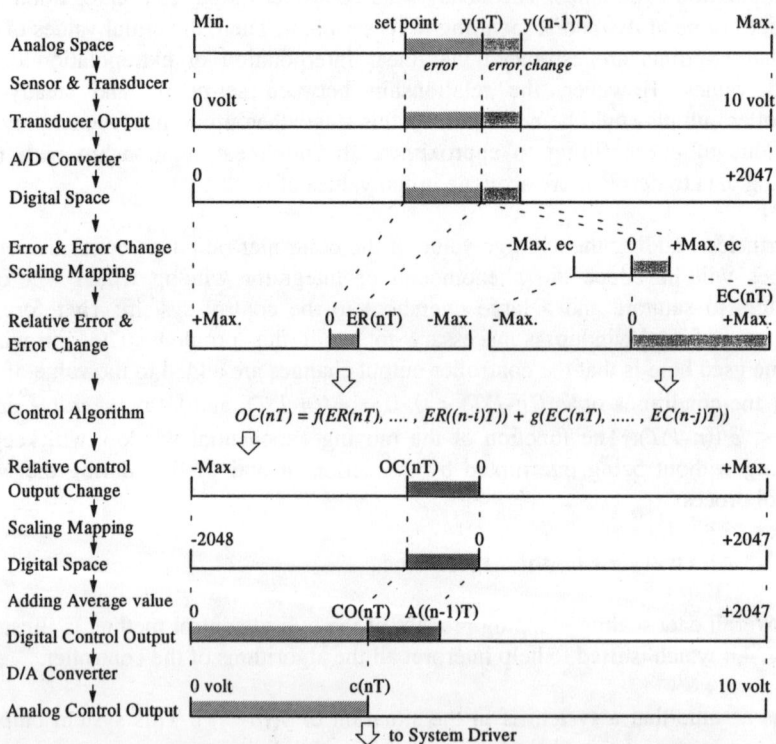

Fig. 4.4   Scaling Mapping Process of Hybrid Qualitative and
              Quantitative Control

The controller output change $OC(nT)$ is generated according to the predicted system states and the system history. Next, $OC(nT)$ is mapped to its corresponding digital output space. The mapped data will be added to the average value of past controller outputs to produce the controller output $CO(nT)$. Then, $CO(nT)$ will be converted to the analogue controller output and sent to the system driver to regulate the system. Ideally, the system will reach the predicted state $y(nT)$ at the next sampling period.

The above has discussed the hybrid qualitative and quantitative control method. Through this method, precise quantitative information and qualitative knowledge representation are integrated effectively. In practical control applications, the system outputs are usually crisp and easily measured, so these quantitative data can be utilised economically. Directly applying quantitative data can avoid the problems caused by mapping quantitative measurement into qualitative space; i.e. how to specify the intervals of the qualitative partitions and what is the optimal number of the partitions. Moreover, since quantitative data can be computed via common operations of real number, no special qualitative operators need to be defined and ambiguities will never occur in the control algorithms. The remaining sections will examine the performance of this hybrid control method via a number of experiments.

4.5   CASE STUDY

In this section, the implementation of the hybrid qualitative and quantitative control method is firstly illustrated by a SISO coupled tanks liquid level control rig. The ways in which the control results are affected by the forgetting factor, anti-windup, system parameters, and the detail-level of bond graph models are investigated. Then, the implementation will be extended to MIMO cases to show that this control method can be applied easily to MIMO control. All the experiments were performed by using Modula-2 programming language, while a graphical man-machine interface was constructed for the purpose of being user-friendly.

4.5.1   SISO Cases

Fig. 4.5 illustrates the schematic diagram of the coupled tanks liquid control rig used in the SISO control experiments. The apparatus consists of two hold-up open tanks coupled by an orifice (R3). Liquid is fed into Tank 1 (C1) by a pump, driven by a continuously variable speed DC motor, at the rate regulated by the control commend. Then, liquid flows into Tank 2 (C2) by means of the liquid pressure in Tank 1. The liquid then flows to the sump through a discharge tap (R4) which can be opened to varying extents.

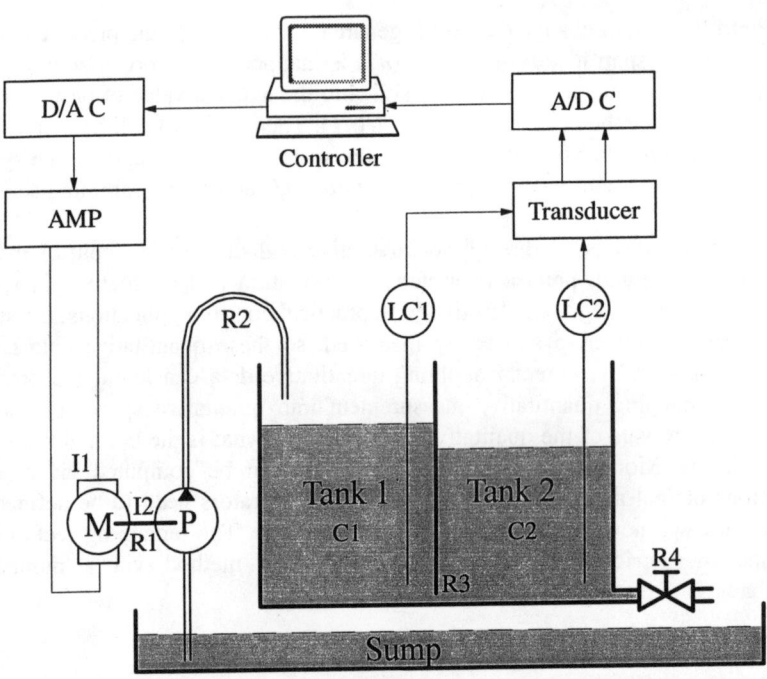

Fig. 4.5 Schematic Diagram of the SISO Coupled Tanks Liquid Level Control Rig

The liquid levels in both tanks are detected by two parallel track depth sensors (LC1 and LC2) placed vertically in each tank. The theme is to control the liquid level of Tank 1 or Tank 2 so as to track the given set-point. Therefore, the liquid level will be the system output, while the controlled voltage used to regulate the DC motor's speed will be the system input. The measurements of liquid levels are converted to digital signals by the transducer and A/D converter, and then sent to the hybrid qualitative and quantitative controller. The digital control commands are transferred to analogue voltage signals by the D/A converter and amplifier. The equipment and their default values set-up used in this implementation are listed in Table 4.1.

Equipment:

Item	Type	QTY.
Controller	IBM PC 286-16	1
A/D D/A Converter	Analog Devices RTI-815 I/O Board	1
Coupled Tanks Rig (Including Transducer, Amplifier, Motor, & Pump)	Tecquipment Coupled Tanks Apparatus CE5	1

Default Value Set-up:

Variable	Value Set-up
A/D Converter:	
Analog input range	0 to +10 V (gain = 1)
Digital output range	0 to +2047
I/O accuracy	±0.02% of full-scale range
D/A Converter:	
Digital input range	0 to +2047
Analog output range	0 to +10 V (gain = 1)
I/O accuracy	±0.02% of full-scale range
Coupled Tanks Rig:	
Liquid level range	0 to 20 cm
Transducer output	0 to +10 V
Amplifier input	0 to +10 V
Sampling Period:	1 sec

Table 4.1 Experimental Equipment and their Default Values Set-up

Nonlinear Behaviour of the Apparatus

There are two obvious nonlinear characteristics in the coupled tanks apparatus. One is the nonlinear measurements caused by the non-linearity of the sensors and the transducer set. Their characteristics are shown in Fig. 4.6.

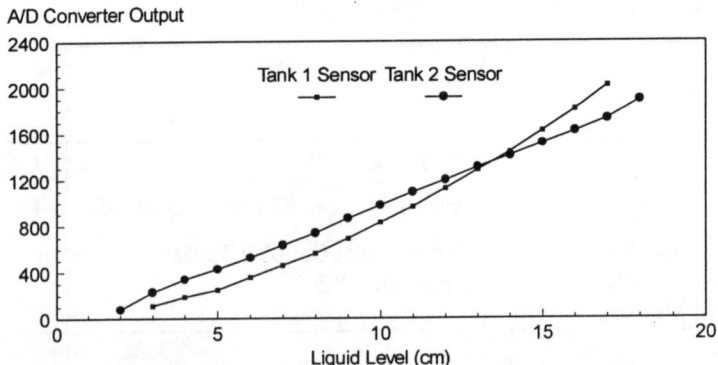

Fig. 4.6 Non-linearity of Liquid Level Sensors

These nonlinear measurements must be correctly calibrated to achieve accurate control results. Accordingly, the technique of curve fitting is employed to produce the regression models for the measurements and A/D conversions:

$$D_1(x_1) = -29.3963 + 29.78064x_1 + 5.87636x_1^2 - 0.03318x_1^3, \qquad (4\text{-}18)$$
$$D_2(x_2) = -141.52572 + 121.17568x_2 - 1.56240x_2^2 + 0.05770x_2^3, \qquad (4\text{-}19)$$

where x_1 is the liquid level of Tank 1 measured in *cm* while D_1 is its corresponding digital output, and x_2 indicates the liquid level of Tank 2 in cm while D_2 is its digital value. A set-point is given in *cm* and transferred through Eq. (4-18) or (4-19) to its corresponding digital value. Thus, system errors and error changes will be obtained by referring to the calibrated digital value of the set-point.

The other non-linearity shown in Fig. 4.7 is caused by the nonlinear transition between the static friction and the dynamic friction of the motor and pump set. When the digital controller output is below 360, the pump will stop pumping. The liquid level will decrease rapidly and lead to a big controller output. The controller output will keep increasing until surpasses than 600. After that, the pump will suddenly start to work and cause the liquid level to increase rapidly. The abnormal increased liquid level will drop the controller output quickly to a value below 360. Then, the pump will again stop pumping. This sequence will reiterate and finally diverge the control system.

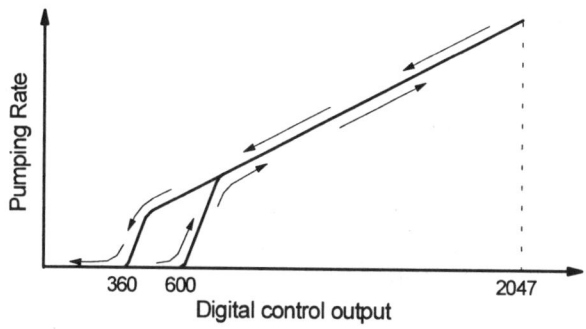

Fig. 4.7 Non-linearity of the DC Motor and Pump Set

A simple strategy is employed to overcome this problem. That is, the output interface will be set to limit the digital controller output to greater than 360 so that the pump will never stop working. Fig. 4.7 shows only three significant values to address the obvious non-linearity. Actually, nonlinear behaviour also appears in the region where the digital controller output is bigger than 600. However, no effort has been made to compensate for this non-linearity. The following experimental results will demonstrate that the hybrid qualitative and quantitative controller can cope with other nonlinear behaviour of the motor and pump.

Derivation of the Control Algorithms

Firstly, the coupled tanks apparatus is modelled via the qualitative bond graph modelling process. Its bond graph model is shown in Fig. 4. 8.

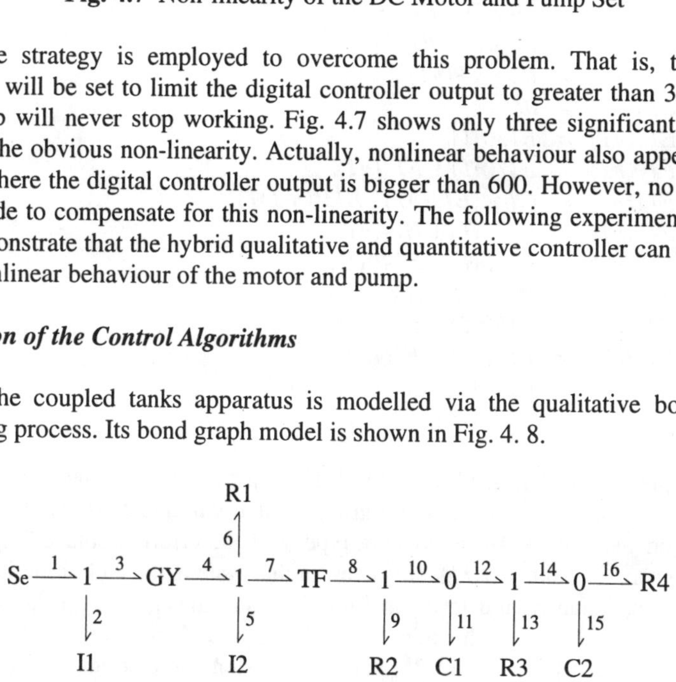

$$Se \xrightarrow{1} 1 \xrightarrow{3} GY \xrightarrow{4} 1 \xrightarrow{7} TF \xrightarrow{8} 1 \xrightarrow{10} 0 \xrightarrow{12} 1 \xrightarrow{14} 0 \xrightarrow{16} R4$$

with R1 at bond 6 above the first 1-junction, I1 at bond 2, I2 at bond 5, R2 at bond 9, C1 at bond 11, R3 at bond 13, C2 at bond 15.

Fig. 4.8 Bond Graph Model of The SISO Coupled Tanks Control Rig

Here, I1 denotes the armature inductance of the motor. I2 denotes the inertia of the motor and pump, while R1 is their axial friction. R2 indicates the flow resistance of the inlet pipe, R3 is the flow resistance of the orifice between the two tanks, and R4 is that of the discharge tap. C1 is the capacitance of Tank 1, while C2 is the capacitance of Tank 2. The qualitative representation of this bond graph model is written as:

$$E1(nT) = E2(nT) + E3(nT), \tag{4-20}$$
$$F1(nT) = F2(nT) = F3(nT), \tag{4-21}$$
$$E2(nT) = I1 \times (F2(nT) - F2((n-1)T), \tag{4-22}$$
$$E3(nT) = F4(nT), \tag{4-23}$$
$$F3(nT) = E4(nT), \tag{4-24}$$
$$E4(nT) = E5(nT) + E6(nT) + E7(nT), \tag{4-25}$$
$$F4(nT) = F5(nT) = F6(nT) = F7(nT), \tag{4-26}$$
$$E5(nT) = I2 \times (F5(nT) - F5((n-1)T), \tag{4-27}$$
$$E6(nT) = R1 \times F6(nT), \tag{4-28}$$
$$E7(nT) = E8(nT), \tag{4-29}$$
$$F7(nT) = F8(nT), \tag{4-30}$$
$$E8(nT) = E9(nT) + E10(nT), \tag{4-31}$$
$$F8(nT) = F9(nT) = F10(nT), \tag{4-32}$$
$$E9(nT) = R2 \times F9(nT), \tag{4-33}$$
$$E10(nT) = 0, \tag{4-34}$$
$$E11(nT) = E12(nT), \tag{4-35}$$
$$F10(nT) = F11(nT) + F12(nT), \tag{4-36}$$
$$F11(nT) = C1 \times (E11(nT) - E11((n-1)T), \tag{4-37}$$
$$E12(nT) = E13(nT) + E14(nT), \tag{4-38}$$
$$F12(nT) = F13(nT) = F14(nT), \tag{4-39}$$
$$E13(nT) = R3 \times F13(nT), \tag{4-40}$$
$$E14(nT) = E15(nT) = E16(nT), \tag{4-41}$$
$$F14(nT) = F15(nT) + F16(nT), \tag{4-42}$$
$$F15(nT) = C2 \times (E15(nT) - E15((n-1)T), \tag{4-43}$$
$$E16(nT) = R4 \times F16(nT). \tag{4-44}$$

In this representation, Eqs. (4-34) and (4-35) express a special interaction between the system variables. From the bond graph model, variable $E10$ denotes the effort working on the exit of the liquid inlet pipe, and the effort should be equal to the liquid pressure in Tank 1 $(E11)$. However, in this system, the end of the inlet pipe is above the maximum liquid level of Tank 1, which means that the liquid level in Tank 1 has no effect on the inlet flow. Therefore, their interaction is specially described in Eqs. (4-34) and (4-35). This expression can distinguish the difference between the end of the inlet pipe above or below the Tank 1 liquid level.

Now, if the control goal is to regulate the liquid level in Tank 2, then the system error can be identified as $E14$, while $F15$ will be the error change and $E1$ will be the system input. Thus, the qualitative equations can be simplified as

$$\begin{aligned} E1(nT) = {}& 12E14(nT) - 20E14((n-1)T) + 11E14((n-2)T) - 2E14((n-3)T) \\ & + 8F15(nT) - 12F15((n-1)T) + 7F15((n-2)T) - F15((n-3)T). \end{aligned} \tag{4-45}$$

Furthermore, the I/O variables in this simplified equation can be replaced by OC_2 (controller output change for Tank 2 liquid level control), ER_2 (error of Tank 2 liquid level) and EC_2 (error change of Tank 2 liquid level) to obtain the control algorithm:

$$OC_2(nT) = 12ER_2(nT) - 20ER_2((n-1)T) + 11ER_2((n-2)T) - 2ER_2((n-3)T)$$
$$+ 8EC_2(nT) - 12EC_2((n-1)T) + 7EC_2((n-2)T) - EC_2((n-3)T). \qquad (4-46)$$

On the other hand, if the control goal is to regulate the liquid level in Tank 1, then the system output variables will be $E11$ and $F11$, while the system input will still be $E1$. Thus, the control algorithm is generated as:

$$OC_1(nT) = 4ER_1(nT) - 4ER_1((n-1)T) + ER_1((n-2)T)$$
$$+ 4EC_1(nT) - 4EC_1((n-1)T) + EC_1((n-2)T). \qquad (4-47)$$

So far the control algorithms of the SISO coupled tanks liquid level control rig have been obtained. On-line control experiments will be performed based on these two control algorithms and the control results are as follows.

Ex. 4.1: *Step Response*

The theme of the following experiments is to control Tank 2 liquid level with the discharge tap of Tank 2 (R4) half opened. The system step responses to different set-point inputs were used to test the performance of this control method. Firstly, the values of error scaling factor (SF_e), error change scaling factor (SF_{ec}), output scaling factor (SF_o), and forgetting factor (δ) were all chosen as 1. Thus, the basic control performance without tuning any factors can be observed. Fig. 4.9 represents the control result, which shows that the controller can regulate the system to track the set-point stably.

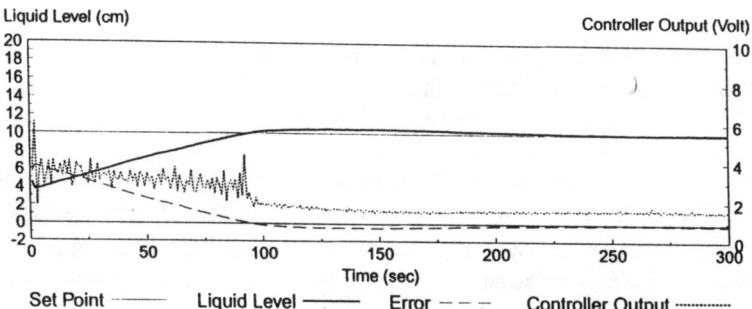

Fig. 4.9 Step Response of Tank 2 Liquid Level;
Set-point = 10 cm, $SF_e = 1$, $SF_{ec} = 1$, $SF_o = 1$, $\delta = 1$.

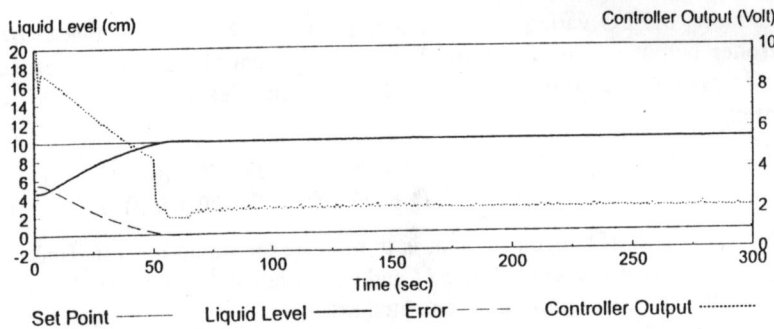

Fig. 4.10 Step Response of Tank 2 Liquid Level;
Set-point = 10 cm, $SF_e = 0.5$, $SF_{ec} = 25$, $SF_o = 0.5$, $\delta = 1$.

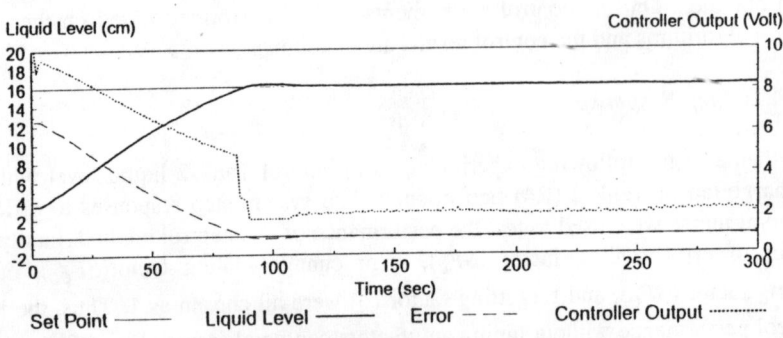

Fig. 4.11 Step Response of Tank 2 Liquid Level;
Set-point = 16 cm, $SF_e = 0.5$, $SF_{ec} = 25$, $SF_o = 0.5$, $\delta = 1$.

However, Fig. 4.9 also shows that the controller output has a serious oscillation in the transient state. This is because that the controller is too sensitive to the system error change. As discussed previously, serious oscillation of controller output is not desired, since it will shorten the life time of actuators. Therefore, the controller's sensitivity to the error change should be reduced. From Eq. (4-4), this requirement can be achieved by increasing the value of SF_{ec}. In this chapter, the choice of scaling factors was refined by trial-and-error, and the most favoured values obtained were $SF_e = 0.5$, $SF_{ec} = 25$, and $SF_o = 0.5$. Figs. 4.10 and 4.11 show the control results with these values. It can be seen that the oscillation of the controller output is eliminated, and the overall control performance is improved: smaller overshoot, shorter rise time and settling time. Moreover, the experimental results also show that the controller can adapt to different set-points without re-tuning scaling factors.

Ex. 4.2: *The Effects of the Techniques Used in the Output Interface*

The experiments performed next were used to investigate the effects of anti-windup and adding controller output change (*OC*) to the average value of past controller outputs (*A*). The control goal and the situation of R4 are the same as in Ex. 4.1. Fig. 4.12 illustrates the control result with controller output changes added to the last controller output rather than the average value of past controller outputs. It can be seen that a cyclic variation occurs in the controller output, while this controller output also causes a small oscillation to the system output. From this example, it was found that adding the average value of previous controller outputs to the controller output changes has a significant smoothing effect on the controller output, and the system output will thus be more stable.

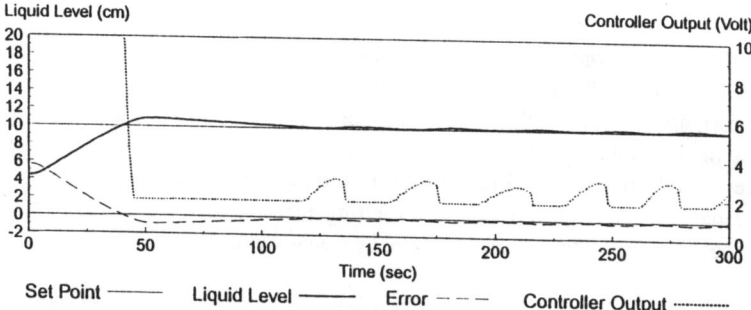

Fig. 4.12 Step Response of Tank 2 Liquid Level;
applying anti-windup, $A(nT) = CO((n-1)T)$,
Set-point = 10 cm, $SF_e = 0.5$, $SF_{ec} = 25$, $SF_O = 0.5$.

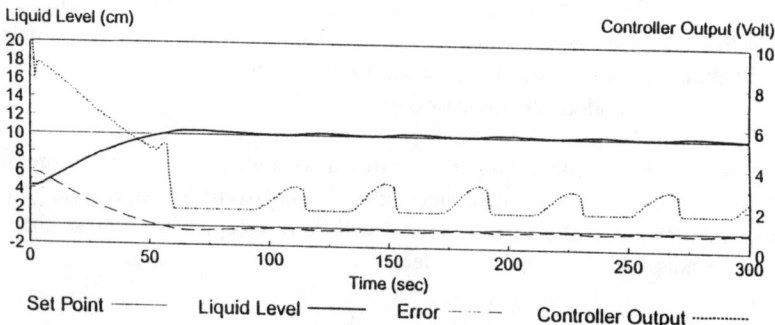

Fig. 4.13 Step Response of Tank 2 Liquid Level;
without applying anti-windup, $A(nT) = CO((n-1)T)$
Set-point = 10 cm, $SF_e = 0.5$, $SF_{ec} = 25$, $SF_O = 0.5$.

Next, the function of anti-windup was switched off to test its effect and the result is shown in Fig. 4.13. An apparent overshoot occurs in the system output, and the controller output is saturated when the system is in the transient state. Comparing this result to Fig. 4.11 and Fig. 4.12, switching off the anti-windup caused a shorter rising time but a longer settling time. Thus, in the applications where short rising time is very important while settling time is less important, the function of anti-windup can be switched off. However, in cases which both rising time and settling time are required to be short, the anti-windup should not be switched off. The requirement can be achieved by using other strategies, such as on-line tuning of the scaling factors. This will be discussed in the next chapter.

Ex. 4.3: *The Effect of Forgetting Factor*

Although adding controller output changes to the average value of previous controller outputs will smooth the controller output, it also stores a large amount of past information which will make it harder for the controller to adapt to rapid changes of the system. Therefore, a moving exponential window is applied to the average value to make it gradually forget past data. The decay rate of the influence-level of past data is adjusted by the forgetting factor δ (see Eq. (4-17)). Smaller δ will cause more rapid decay for past data, and vice versa.

The following experiments were applied to verify the effect of δ. Here, the set-point of the liquid level in Tank 2 was changed from 10 cm to 11 cm at 250 second and reduced back to 10 cm at 450 second to test the adaptability of the controller. Fig. 4.14 to Fig. 4.16 show the control results performed with different δ (1, 0.5, and 0.1). It can be seen that smaller δ makes the controller track the set-point changes better but also causes the controller output and system output to oscillate. It is difficult to determine the best value for δ, and its value should be chosen in terms of the individual needs of various applications. For this coupled tanks liquid level control rig, the most adequate value of δ was 0.9.

These control results imply that the control adaptability cannot be improved by merely decreasing the value of δ, since it will bring some negative effects to the system. A better automatic tuning method will be developed in the next chapter to enhance the adaptability of the controllers.

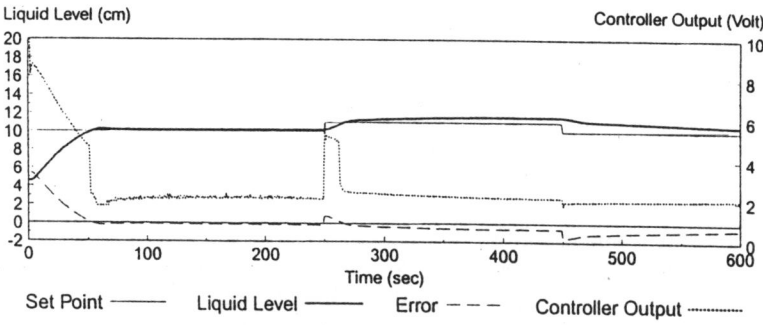

Fig. 4.14 Step Response of Tank 2 Liquid Level; Set-point changed at 250 sec and 450 sec, $SF_e = 0.5$, $SF_{ec} = 25$, $SF_o = 0.5$, $\delta = 1$.

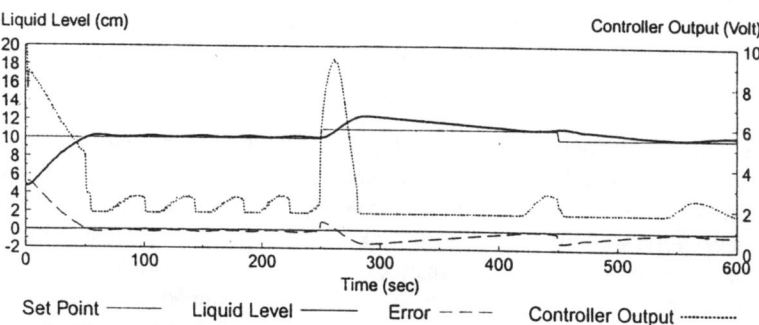

Fig. 4.15 Step Response of Tank 2 Liquid Level; Set-point changed at 250 sec and 450 sec, $SF_e = 0.5$, $SF_{ec} = 25$, $SF_o = 0.5$, $\delta = 0.5$.

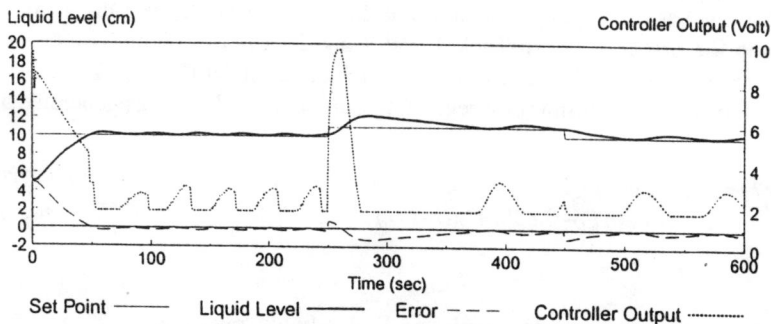

Fig. 4.16 Step Response of Tank 2 Liquid Level; Set-point changed at 250 sec and 450 sec, $SF_e = 0.5$, $SF_{ec} = 25$, $SF_o = 0.5$, $\delta = 0.1$.

Ex. 4.4: *The Influences of Model-Detail-Level and Introducing System Parameters*

So far the control algorithm used in above experiments is derived directly from a bond graph model without considering the system parameters. This control algorithm is decided at the modelling stage according to the system structure. Thus, besides tuning the scaling factors, adjusting the system model to generate different control algorithms will also affect the control performance. There are two occasions which will incur different control algorithms for a system: one is introducing system parameters to the qualitative model and the other is changing the detail-level for the qualitative model. Introducing system parameters modifies the weights of system output variables for a control algorithm, while changing the detail-level of a qualitative bond graph model entirely affects the complexity and completeness of the control algorithm. It can be inferred that both the actions have strong effects on the control performance. In the following experiments, a set of estimated parameters will be inserted into the qualitative model of the coupled tanks liquid level control rig; furthermore this rig was also re-modelled at a lower detail-level, to test their influences.

In the coupled tanks liquid level control rig, some system parameters, such as the inertia of the motor and the friction of the pump, are not easy to evaluate. Therefore, the parameters used here were estimated via experimental experience. Firstly, the capacitances (cross-sectional areas) of Tank 1 and Tank 2 were regarded as being very important for the system dynamics, so that the parameter C1 (Tank 1 capacitance) and C2 (Tank 2 capacitance) were both assumed as 1. Secondly, the flow resistance of Tank 2 discharge tap (R4) and the axial friction of the pump (R1) were regarded as having big values and having great effects on the system, so they were also assumed as 1. Next, the armature inductance of the motor (I1) and the inertia of the pump and motor (I2) in this system were relative small, so their values were assumed as 0.1. Finally, the values of the flow resistances of Tank 1 inlet pipe (R2) and the orifice between Tank 1 and Tank 2 (R3) were both assume as 0.5. Then, these values were inserted into the qualitative model (Eqs. (4-20) to (4-44)) and thus the control algorithm for regulating Tank 2 liquid level was generated as:

$$OC_2(nT) = 2.9ER_2(nT) - 2.165ER_2((n\text{-}1)T) + 0.28ER_2((n\text{-}2)T) - 0.015ER_2((n\text{-}3)T) + $$
$$1.74EC_2(nT) - 0.835EC_2((n\text{-}1)T) + 0.1EC_2((n\text{-}2)T) - 0.005EC_2((n\text{-}3)T),$$
$$(4\text{-}48)$$

while the control algorithm for regulating Tank 1 liquid level was:

$$OC_1(nT) = 2.32ER_1(nT) - 0.34ER_1((n-1)T) + 0.02ER_1((n-2)T) + 1.16EC_1(nT) -$$
$$0.17EC_1((n-1)T) + 0.01EC_1((n-2)T). \tag{4-49}$$

It can be seen that the interrelations of the weights of the system output variables is quite different from that of the control algorithms generated without considering system parameters (Eqs. (4-46) and (4-47)).

On the other hand, the coupled tanks rig was re-modelled at a lower detail-level without considering the system parameters. Here, the primitives I1 and I2 were ignored, since their effects on the system were considered to be relatively small compared with other components. Fig. 4.17 shows the simplified bond graph model.

$$Se \xrightarrow{1} GY \xrightarrow{2} 1 \xrightarrow{4} TF \xrightarrow{5} 1 \xrightarrow{7} 0 \xrightarrow{9} 1 \xrightarrow{11} 0 \xrightarrow{13} R4$$

with downward bonds: $\downarrow 3$ to R1, $\downarrow 6$ to R2, $\downarrow 8$ to C1, $\downarrow 10$ to R3, $\downarrow 12$ to C2

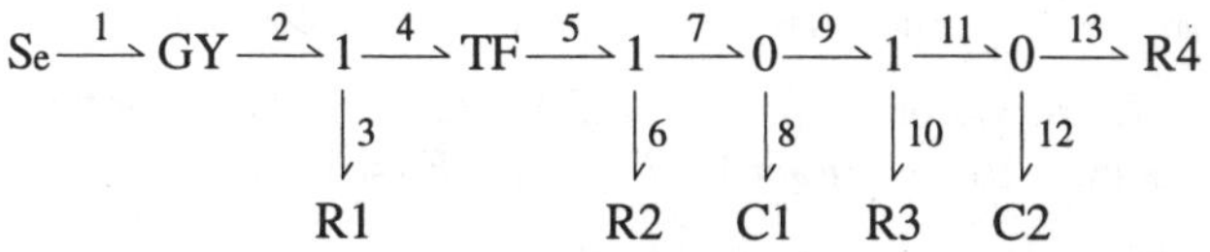

Fig. 4.17 Simplified Bond Graph Model of the SISO Coupled Tanks Control Rig

Based on this model, the control algorithms used to control Tank 2 and Tank 1 liquid levels are derived respectively as Eqs. (4-50) and (4-51).

$$OC_2(nT) = 3ER_2(nT) - 2ER_2((n-1)T) + 2EC_2(nT) - EC_2((n-1)T), \tag{4-50}$$
$$OC_1(nT) = ER_1(nT) + EC_1(nT). \tag{4-51}$$

The control algorithms given in Eqs. (4-46), (4-48), and (4-50) were applied respectively to regulate the Tank 2 liquid level with the discharge tap of Tank 2 (R4) half opened, while the control algorithms Eqs. (4-47), (4-49), and (4-51) were used to control the Tank 1 liquid level with R4 also half opened. Firstly, the scaling factors and the forgetting factor were all given by 1. The control result using the control algorithm Eq. (4-46) has already been shown in Fig. 4.9 and the results of using Eqs. (4-48) and (4-50) are shown in Figs. 4.18 and 4.19 respectively. It can be seen from comparing Figs. 4.9 and 4.18 that the control algorithm which is derived from the model with the system parameters involved provides a better control performance — shorter rise time and settling time. The reason for this is that inserting the system parameters into the qualitative model makes the model more accurate. Theoretically, a more accurate model can derive a more adequate control algorithm for the system. Besides, the control algorithm derived from the simplified model also provides a better control performance (Fig. 4.19) than that performed by the complex model (Fig. 4.9). The reason is that the simplified model is much closer to the accurate one in the relation to the variable weights than that of the complex model. This is also the reason why Figs. 4.18 and 4.19 are so similar.

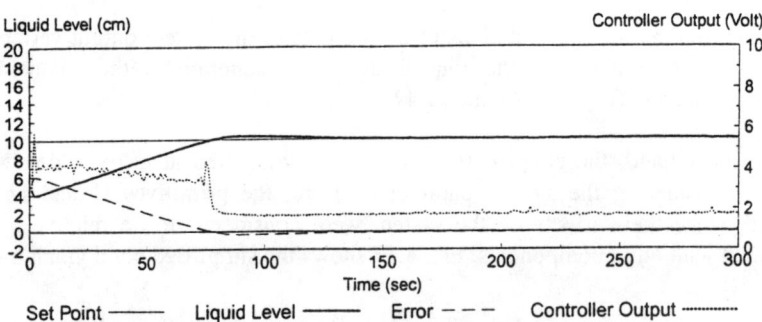

Fig. 4.18 Step Response of Tank 2 Liquid Level; Control Algorithm: Eq. (4-48), Set-point = 10 cm, $SF_e = 1$, $SF_{ec} = 1$, $SF_o = 1$,

$\delta = 1$.

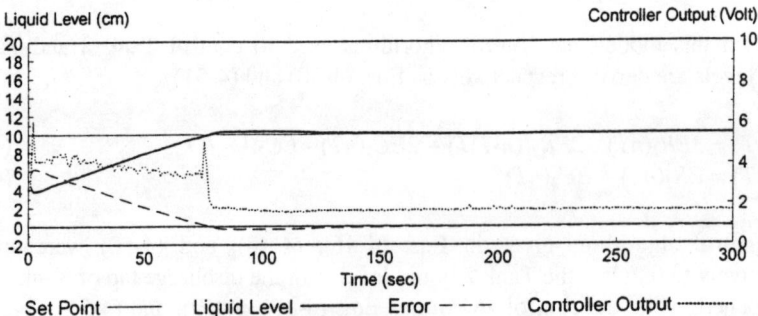

Fig. 4.19 Step Response of Tank 2 Liquid Level; Control Algorithm: Eq. (4-50), Set-point = 10 cm, $SF_e = 1$, $SF_{ec} = 1$, $SF_o = 1$,

$\delta = 1$.

Next, the favoured values of the scaling factors and forgetting factor for this system were applied to the controllers to regulate Tank 2 and Tank 1 liquid levels respectively. The control results are shown in Figs. 4.20 to 4.25. These results lead to the conclusion that the control algorithms derived from the accurate and simplified models provide a better control performance. It also can be seen from comparing Figs. 4.9, 4.18, 4.19, and 4.20 that tuning the scaling factors can improve the control performance more effectively than adjusting the system model. For example, the controller output oscillation in the transient state cannot be eliminated by adjusting the model but can be eliminated by tuning the scaling factors. This is to say that some disadvantages caused by the lack of taking the system parameters into account can be compensated by tuning the scaling factors. Therefore, in this control methodology, it is not necessary to evaluate system parameters for a satisfactory control performance.

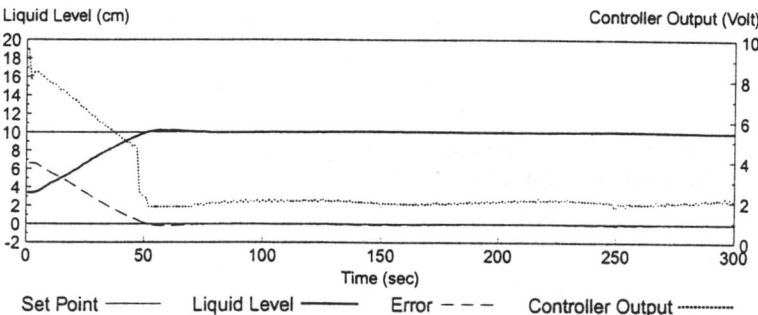

Fig. 4.20 Step Response of Tank 2 Liquid Level; Control Algorithm: Eq. (4-46),
Set-point = 10 cm, $SF_e = 0.5$, $SF_{ec} = 25$,
$SF_o = 0.5$, $\delta = 0.9$.

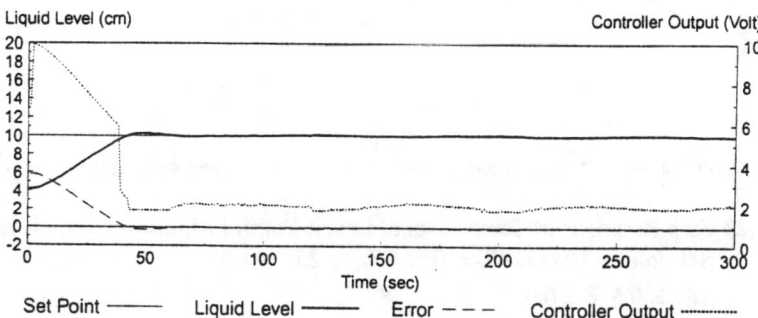

Fig. 4.21 Step Response of Tank 2 Liquid Level; Control Algorithm: Eq. (4-48),
Set-point = 10 cm, $SF_e = 0.5$, $SF_{ec} = 25$,
$SF_o = 0.5$, $\delta = 0.9$.

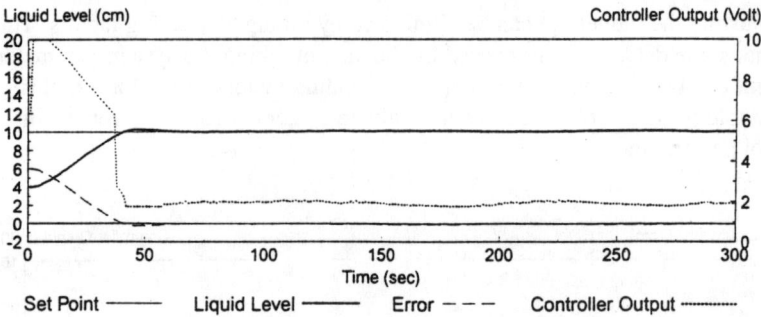

Fig. 4.22 Step Response of Tank 2 Liquid Level; Control Algorithm: Eq. (4-50),
Set-point = 10 cm, $SF_e = 0.5$, $SF_{ec} = 25$,
$SF_o = 0.5$, $\delta = 0.9$.

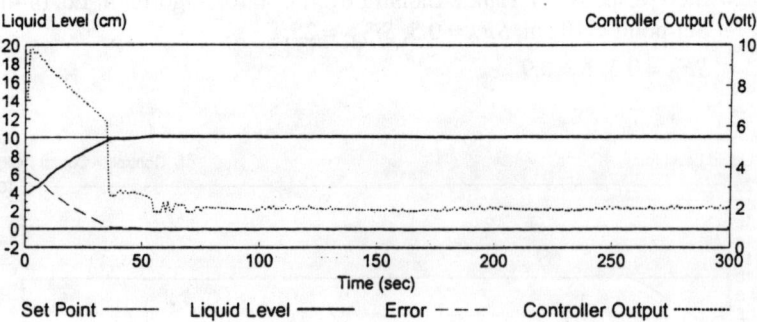

Fig. 4.23 Step Response of Tank 1 Liquid Level; Control Algorithm: Eq. (4-47),
Set-point = 10 cm, $SF_e = 0.5$, $SF_{ec} = 25$,
$SF_o = 0.5$, $\delta = 0.9$.

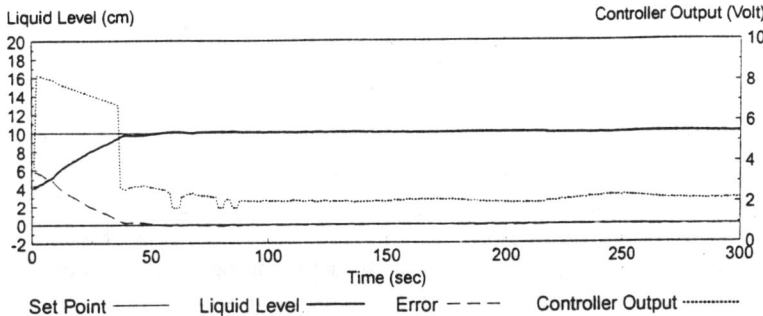

Fig. 4.24 Step Response of Tank 1 Liquid Level; Control Algorithm: Eq. (4-49), Set-point = 10 cm, $SF_e = 0.5$, $SF_{ec} = 25$, $SF_O = 0.5$, $\delta = 0.9$.

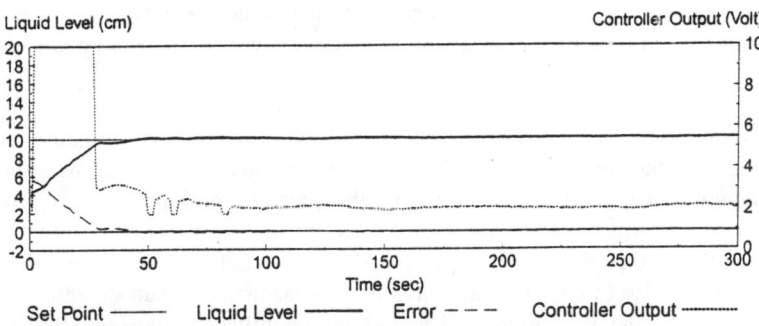

Fig. 4.25 Step Response of Tank 1 Liquid Level; Control Algorithm: Eq. (4-51), Set-point = 10 cm, $SF_e = 0.5$, $SF_{ec} = 25$, $SF_O = 0.5$, $\delta = 0.9$.

The experimental results also suggest that the complex model used to derive the control algorithms Eqs. (4-46) and (4-47) is not the most suitable one for the control system. In this system, the effects of I1 and I2 are small enough to be ignored. However, the complex model involves I1 and I2 but does not employ the parameters to denote that they are less important than other primitives. Therefore, this complex model cannot describe the system behaviour very accurately and thus the control algorithms derived from it are not the best ones. This implies that when the consideration of system parameters is not involved in a qualitative modelling process, the detail-level will decide the accuracy for a qualitative model. If tuning scaling factors still cannot fulfil the performance requirements, then adjusting the qualitative model to be more accurate provides a further possibility to generate control algorithms to meet the performance criteria.

Let us now consider the control algorithm described in Eq. (4-51). This control algorithm is the simplest one in this control method, since it contains only one error and one error change variable. In practice, this control strategy is very close to the PI control method. Here, the output scaling factor can be seen as the proportional feedback gain, and the error scaling factor plays a role similar to the integral item, yet the error change scaling factor has no corresponding parameter in PI controllers. From this discussion, the hybrid control methodology can be seen as an enhanced PI control method. However, in contrast to conventional PI controllers, the hybrid controllers can be designed without any numerical details about the controlled systems. In addition, the control algorithms used in the hybrid control method contain historical information of system behaviour so that these controllers can regulate more stably the higher-order systems within wider operation ranges without re-tuning the scaling factors. A further difference which will be discussed in the following section is that the hybrid control method is very easily extended to control MIMO systems.

Although the accurate and simple bond graph models produced effective control algorithms for the coupled tanks apparatus, the complex one will still be applied in the following chapters. One reason is that the control algorithms derived from the complex model will be used to test the capability of the auto-tuning method. The other reason is that the bond graph model is used not only to generate control algorithms but also to construct the inference mechanism for fault diagnosis. At the fault diagnosis stage, the variables I1 and I2 will be utilised to represent the faults of the motor and pump, so they are kept in the bond graph model.

4.5.2   MIMO Cases

Fig. 4.26 shows the schematic diagram of the MIMO coupled tanks liquid level control rig. Another set of amplifier, motor, and pump is appended to the SISO coupled tanks apparatus to pump liquid directly into Tank 2. Thus, the liquid level in Tank 2 will not only relate to the liquid level in Tank 1 but also depend on the flow fed by the second pump. The theme of the MIMO control case is to control the liquid levels of Tank 1 and Tank 2 to reach their own set-points by tuning the pumping rates of the two pumps.

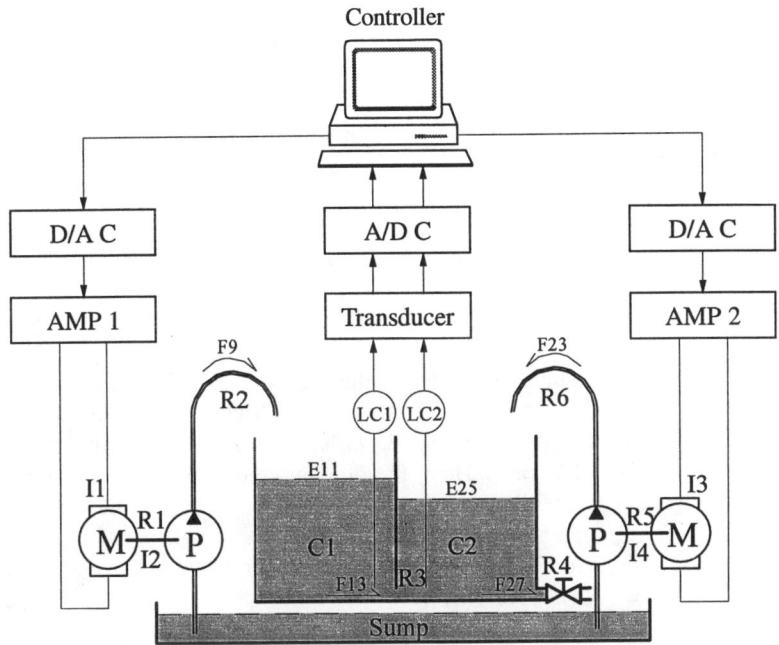

Fig. 4.26 Schematic Diagram of the MIMO Coupled Tanks Liquid Level Control Rig

The type of motor and pump appended to the SISO coupled tanks apparatus was Mini Puppy Model 8860-1203 manufactured by ITT Jabsco CO. The amplifier used to drive the second motor was made by the Control Eng. Department of Sheffield Univ. with the same specification as the previous one. All other components and their value set-up are the same as described in Table 4.1.

The sensors used here are also the same as the previous ones, so that Eqs. (4-18) and (4-19) can be employed to compensate for their non-linearity. On the other hand, the second pump can keep up a very small flow quantity without stop pumping when the digital controller output is decreased to zero. Therefore, the strategy which was used to overcome the non-linearity of the first pump is not applied to the second pump.

Derivation of the Control Algorithms

Like the former case, the MIMO coupled tanks rig was firstly modelled by the qualitative bond graph modelling method, and its bond graph model is shown in Fig. 4.27.

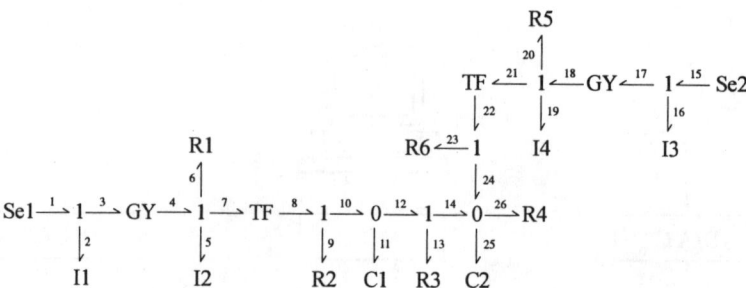

Fig. 4.27 Bond Graph Model of the MIMO Coupled Tanks Control Rig

Here, I3 denotes the armature inductance of the second motor. I4 denotes the inertia of the second motor and pump, while R5 is their axial friction. R6 indicates the flow resistance of the inlet pipe of Tank 2. Other parameters indicate the same components as in the SISO case. Then, the qualitative equations used to represent this model are generated as follows:

$$
\begin{aligned}
E1(nT) &= E2(nT) + E3(nT), & (4\text{-}52) \\
F1(nT) &= F2(nT) = F3(nT), & (4\text{-}53) \\
E2(nT) &= I1 \times (F2(nT) - F2((n\text{-}1)T), & (4\text{-}54) \\
E3(nT) &= F4(nT), & (4\text{-}55) \\
F3(nT) &= E4(nT), & (4\text{-}56) \\
E4(nT) &= E5(nT) + E6(nT) + E7(nT), & (4\text{-}57) \\
F4(nT) &= F5(nT) = F6(nT) = F7(nT), & (4\text{-}58) \\
E5(nT) &= I2 \times (F5(nT) - F5((n\text{-}1)T), & (4\text{-}59) \\
E6(nT) &= R1 \times F6(nT), & (4\text{-}60) \\
E7(nT) &= E8(nT), & (4\text{-}61) \\
F7(nT) &= F8(nT), & (4\text{-}62) \\
E8(nT) &= E9(nT) + E10(nT), & (4\text{-}63) \\
F8(nT) &= F9(nT) = F10(nT), & (4\text{-}64) \\
E9(nT) &= R2 \times F9(nT), & (4\text{-}65) \\
E10(nT) &= 0, & (4\text{-}66) \\
E11(nT) &= E12(nT), & (4\text{-}67) \\
F10(nT) &= F11(nT) + F12(nT), & (4\text{-}68) \\
F11(nT) &= C1 \times (E11(nT) - E11((n\text{-}1)T), & (4\text{-}69) \\
E12(nT) &= E13(nT) + E14(nT), & (4\text{-}70) \\
F12(nT) &= F13(nT) = F14(nT), & (4\text{-}71) \\
E13(nT) &= R3 \times F13(nT), & (4\text{-}72) \\
E15(nT) &= E16(nT) + E17(nT), & (4\text{-}73) \\
F15(nT) &= F16(nT) = F17(nT), & (4\text{-}74) \\
E16(nT) &= I3 \times (F16(nT) - F16((n\text{-}1)T), & (4\text{-}75)
\end{aligned}
$$

$$E17(nT) = F18(nT), \tag{4-76}$$
$$F17(nT) = E18(nT), \tag{4-77}$$
$$E18(nT) = E19(nT) + E20(nT) + E21(nT), \tag{4-78}$$
$$F18(nT) = F19(nT) = F20(nT) = F21(nT), \tag{4-79}$$
$$E19(nT) = I4 \times (F19(nT) - F19((n-1)T), \tag{4-80}$$
$$E20(nT) = R5 \times F20(nT), \tag{4-81}$$
$$E21(nT) = E22(nT), \tag{4-82}$$
$$F21(nT) = F22(nT), \tag{4-83}$$
$$E22(nT) = E23(nT) + E24(nT), \tag{4-84}$$
$$F22(nT) = F23(nT) = F24(nT), \tag{4-85}$$
$$E23(nT) = R6 \times F23(nT), \tag{4-86}$$
$$E24(nT) = 0, \tag{4-87}$$
$$E14(nT) = E25(nT) = E26(nT), \tag{4-88}$$
$$F14(nT) + F24(nT) = F25(nT) + F26(nT), \tag{4-89}$$
$$F25(nT) = C2 \times (E25(nT) - E25((n-1)T), \tag{4-90}$$
$$E26(nT) = R4 \times F26(nT), \tag{4-91}$$

It can be seen from Eq. (4-89) that Tank 2 has two flow sources: one comes from Tank 1 ($F14$) and the other comes from its inlet pipe ($F24$).

According to the control goal, the system output variables can be identified as $E11$ (liquid level in Tank 1), $F11$ (liquid level change in Tank 1), $E14$(liquid level in Tank 2), and $F25$(liquid level change in Tank 2), while the variables of system inputs are $E1$ (voltage input of the first motor) and $E15$ (voltage input of the second motor). With this identification, the qualitative equations can be simplified as follows:

$$
\begin{aligned}
E1(nT) = {}& 4E11(nT) - 4E11((n-1)T) + E11((n-2)T) \\
& + 4F11(nT) - 4F11((n-1)T) + F11((n-2)T) \\
& - 4E14(nT) + 4E14((n-1)T) - E14((n-2)T),
\end{aligned} \tag{4-92}
$$

$$
\begin{aligned}
E15(nT) = {}& 8E14(nT) - 8E14((n-1)T) + 2E14((n-2)T) \\
& + 4F25(nT) - 4F25((n-1)T) + F25((n-2)T) \\
& - 4E11(nT) + 4E11((n-1)T) - E11((n-2)T).
\end{aligned} \tag{4-93}
$$

Then, the I/O variables in these simplified equations can be replaced by the variables of OC_1 (controller output change of the first amplifier), OC_2 (controller output change of the second amplifier), ER_1 (error in Tank 1), ER_2 (error in Tank 2), EC_1 (error change in Tank 1), and EC_2 (error change in Tank 2). Thus, the MIMO control algorithm is derived as follows:

$$OC_1(nT) = 4ER_1(nT) - 4ER_1((n-1)T) + ER_1((n-2)T)$$
$$+ 4EC_1(nT) - 4EC_1((n-1)T) + EC_1((n-2)T)$$
$$- 4ER_2(nT) + 4ER_2((n-1)T) - ER_2((n-2)T), \qquad (4\text{-}94)$$

$$OC_2(nT) = 8ER_2(nT) - 8ER_2((n-1)T) + 2ER_2((n-2)T)$$
$$+ 4EC_2(nT) - 4EC_2((n-1)T) + EC_2((n-2)T)$$
$$- 4ER_1(nT) + 4ER_1((n-1)T) - ER_1((n-2)T). \qquad (4\text{-}95)$$

This MIMO control algorithm is also derived from the system structure by the generalised procedure developed in Chapter 3. For feedback control systems, identifying system I/O variables is the key to producing individual control algorithms for various control goals. After the I/O variables have been decided, other variables which are not chosen as I/O ones will be replaced by the I/O variables according to their interactions. The number of I/O variables makes no difference to the simplification procedure of the qualitative equations. This means that generating MIMO control algorithms will not be more difficult than generating SISO ones.

Further, the anti-windup scheme used here is that the controller output changes of each controller are added by their own initial value of A under the conditions of $\hat{e}c_1((n-1)T) < 0$, $\hat{e}c_2((n-1)T) < 0$, $0 < \hat{e}_1((n-1)T)$, $0 < \hat{e}_2((n-1)T)$, |*Max. negative* $\hat{e}c_1((n-1)T)$| $< \hat{e}_1((n-1)T)$, and |*Max. negative* $\hat{e}c_2((n-1)T)$| $< \hat{e}_2((n-1)T)$. That is, both the controllers will "turn off" the anti-windup function as soon as any of the conditions are violated.

Ex. 4.5: *Step Responses of the MIMO Controller*

Fig. 4.28 to Fig. 4.31 illustrate the step responses of the MIMO controller operating at different set-points. The discharge tap of Tank 2 (R4) was fully opened.

The control results show that the hybrid qualitative and quantitative control method can regulate the MIMO system successfully. In the cases of Fig. 4.28 and Fig. 4.29, the controller outputs of the second controller were very small, which means the distance between the liquid levels in Tank 1 and Tank 2 was very close to that in the SISO cases. This distance cannot be further extended in this coupled tanks rig. The reason is that extending liquid level distance will increase the flow quantity passing through the orifice from Tank 1 to Tank 2. However, the capacity of the discharge tap of Tank 2 is not big enough to drain the extra quantity of the increased flow. Thus, the liquid level in Tank 2 will rise to reduce the liquid level distance until the quantities flowing in and out of Tank 2 balance. That is to say the system structure

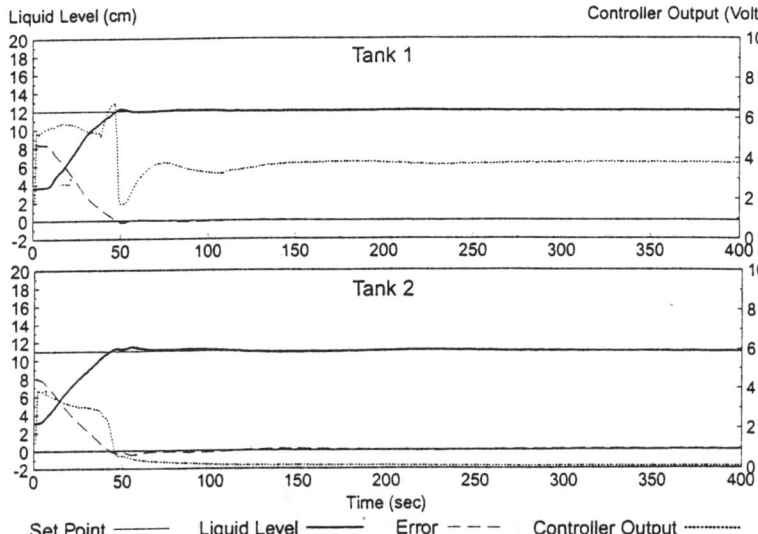

Fig. 4.28 Step Responses of the MIMO Control System;
Tank 1 Set-point = 12 cm, Tank 2 Set-point = 11 cm,
Controller 1: $SF_e = 0.5$, $SF_{ec} = 30$, $SF_o = 0.5$, $\delta = 0.9$,
Controller 2: $SF_e = 0.5$, $SF_{ec} = 30$, $SF_o = 1.0$, $\delta = 0.9$.

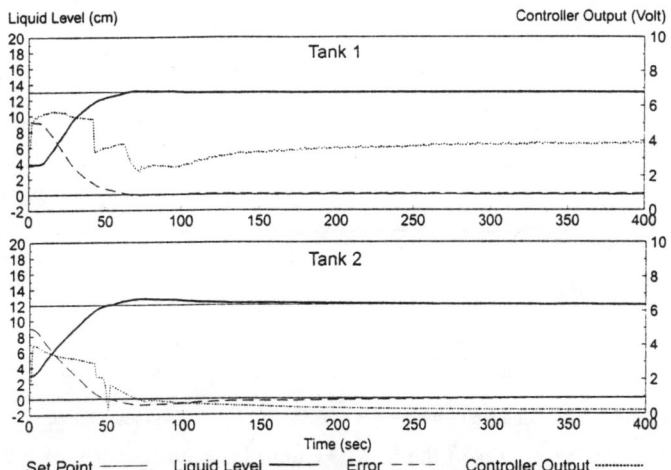

Fig. 4.29 Step Responses of the MIMO Control System;
Tank 1 Set-point = 13 cm, Tank 2 Set-point = 12 cm,
Controller 1: $SF_e = 0.5$, $SF_{ec} = 30$, $SF_o = 0.5$, $\delta = 0.9$,
Controller 2: $SF_e = 0.5$, $SF_{ec} = 30$, $SF_o = 1.0$, $\delta = 0.9$.

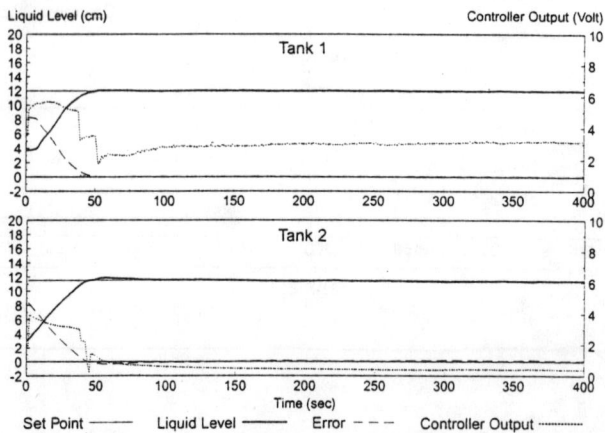

Fig. 4.30 Step Responses of the MIMO Control System;
Tank 1 Set-point = 12 cm, Tank 2 Set-point = 11.5 cm,
Controller 1: $SF_e = 0.5$, $SF_{ec} = 30$, $SF_o = 0.5$, $\delta = 0.9$,
Controller 2: $SF_e = 0.5$, $SF_{ec} = 30$, $SF_o = 1.0$, $\delta = 0.9$.

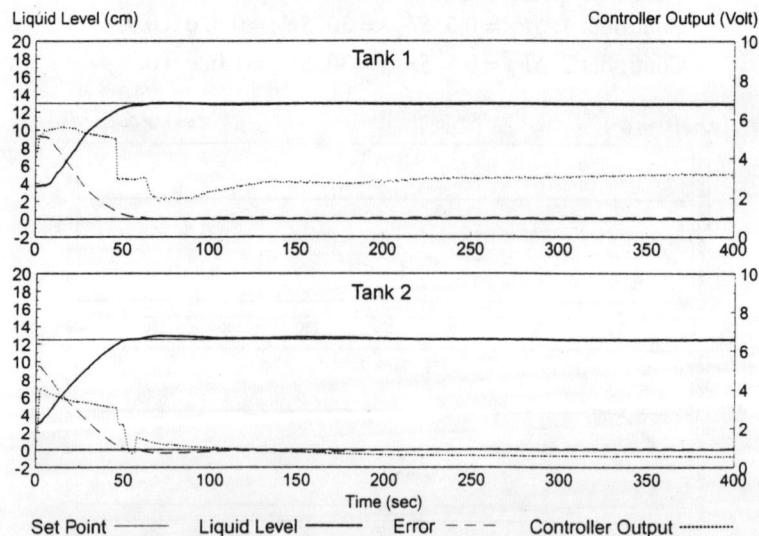

Fig. 4.31 Step Responses of the MIMO Control System;
Tank 1 Set-point = 13 cm, Tank 2 Set-point = 12.5 cm,
Controller 1: $SF_e = 0.5$, $SF_{ec} = 30$, $SF_o = 0.5$, $\delta = 0.9$,
Controller 2: $SF_e = 0.5$, $SF_{ec} = 30$, $SF_o = 1.0$, $\delta = 0.9$.

limits the system behaviour and no controller can regulate the system to overcome the structural limit. The maximum liquid level distance in this system is around 1 cm.

Ex. 4.6: *Adaptability of the MIMO Controller*

In these experiments, the MIMO controller was required to adapt to system parameter changes. Fig. 4.32 shows the result of the case where Tank 2 leaked at 205 second, and Fig. 4.33 illustrates another result where the discharge tap blocked at 300 second.

It can be seen from the results that the controller can adapt to system parameter changes and regulate the system at different set-points without re-tuning the scaling factors. The reason for the existence of this robustness is as discussed in Ex. 4.4, namely that the hybrid qualitative and quantitative controller can regulate a system without considering system parameters. Therefore, parameter changes have no significant influence on the control system. However, in some particular cases, the control algorithms will appear incorrect for the system and cannot regulate the system successfully. For example, when the discharge tap is totally blocked; such a

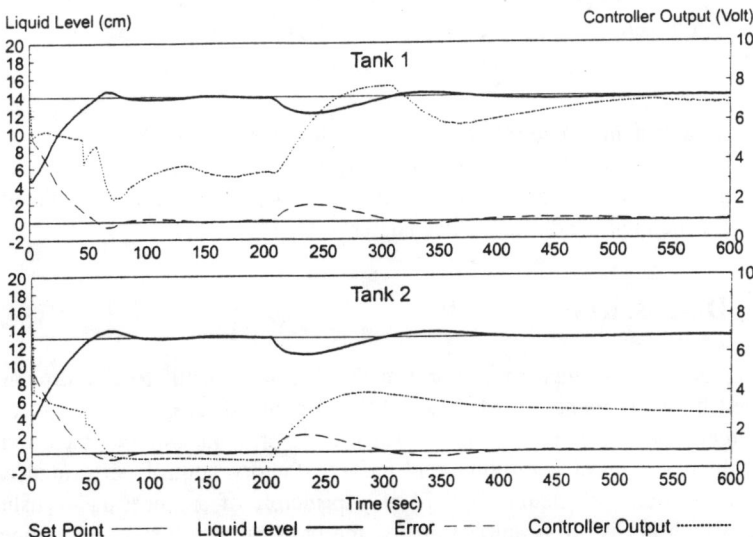

Fig. 4.32 Step Responses of the MIMO Controller; Tank 2 leaks at 205 sec, Tank 1
Set-point = 14 cm, Tank 2 Set-point = 13 cm,
Controller 1: $SF_e = 0.5$, $SF_{ec} = 30$, $SF_o = 0.5$, $\delta = 0.9$,
Controller 2: $SF_e = 0.5$, $SF_{ec} = 30$, $SF_o = 1.0$, $\delta = 0.9$.

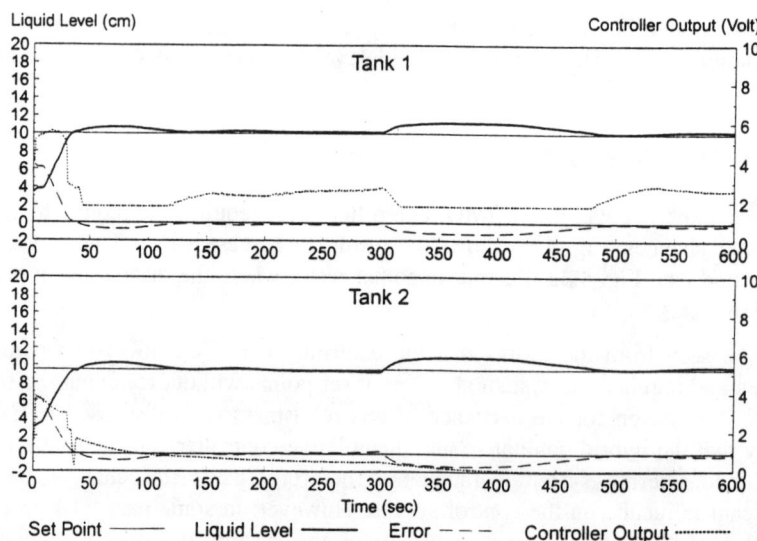

Fig. 4.33 Step Responses of the MIMO Controller; Discharge tap blocks at 300 sec,
Tank 1 Set-point = 10 cm, Tank 2 Set-point = 9.5 cm,
Controller 1: $SF_e = 0.5$, $SF_{ec} = 30$, $SF_O = 0.5$, $\delta = 0.9$,
Controller 2: $SF_e = 0.5$, $SF_{ec} = 30$, $SF_O = 1.0$, $\delta = 0.9$.

situation is a system structural change rather than a parameter change, so that the
control algorithms derived in terms of the system structure are no longer correct.
This kind of structural change will lead the control problem into the field of fault
diagnosis, which will be discussed in Chapter 6.

4.6   DISCUSSION

This chapter has implemented a hybrid qualitative and quantitative control method
illustrated by experiments on SISO and MIMO coupled tanks liquid level control
rigs in real-time. This method successfully integrates quantitative measurements
with qualitative control algorithms and has satisfactory control performance. The
essence of this control method is built on a principle of engineering — using the
material which is easy to acquire. Usually, the physical structure of an engineering
system, such as a pipe connected to a tank or a motor linked with a gear box, is easy
to be understood by controller designers. In contrast, detailed specification, i.e. the
capacity and flow resistance of a pipe or the inertia of a gear box, is relatively
difficult to be evaluated. Consequently, at the controller design stage, it is beneficial
to employ only structural information for constructing controllers with the numerical

details being ignored. Thus, the controller built from this designing process is qualitative. On the other hand, measuring system outputs on-line and computing quantitative data to derive precise control commands are usually easy. However, mapping quantitative measurement into qualitative space and generating explicit control actions from qualitative information need to overcome a lot of difficulties, such as choosing qualitative scales and avoiding the ambiguities caused by qualitative operations. It is profitable to exercise directly the quantitative measurement to produce accurate control commands at the on-line control stage. Thus, the derivation of control actions is quantitative. All these considerations motivate the development of the hybrid qualitative and quantitative control method.

In order to merge quantitative information into qualitative control algorithms, a scaled mapping was developed in this chapter to transfer data between practical quantitative measurements and their relative quantities computed in control algorithms. Moreover, a one step prediction method was proposed to derive control commands with the mapped quantitative information and the knowledge represented in qualitative control algorithms. The knowledge representation proposed in Chapter 3 has a mathematical form so that it can naturally deal with the quantitative data and support producing control commands via conventional mathematical operations.

From the implementation of this control method, some peculiarities compared to conventional control methods are as follows:

- The control algorithms can be generated from a system structure through a generalised procedure without any numerical details.

- The controller designing process is generally suitable for both SISO and MIMO tasks.

- The hybrid controllers are less sensitive to system parameter changes, since their control algorithms contain no parameter information.

Compared to existing qualitative control methods, this methodology also has the following unique features:

- The employment of quantitative information enables the qualitative control algorithms to derive accurate control commands.

- The usage of conventional mathematical operations saves computing time for the derivation of control commands, which is very helpful for executing on-line control.

Nevertheless, there remain three problems in this control method, which need further consideration. Firstly, this control method employs scaling factors to adjust the control performance. It distracts the controller design away from the specifics of an application and offers the controller freedom to tune its response to individual application needs. That is to say the choice of scaling factors will affect the success of a controller. However, there seems no general computational method for predicting optimum scaling factors, since there is no numerical information employed in the qualitative controller design process to support this prediction. A possible solution to this problem is automatically tuning the scaling factors on-line according to the observation of system behaviour. This will be discussed in Chapter 5.

Secondly, as discussed in Ex. 4.4, the detail-level of qualitative bond graph models determines the formulation of control algorithms, and it is also one of the essential factors affecting the control performance. Again, the optimum detail-level is hard to predict through qualitative reasoning methods, while on-line adjusting the modelling detail-level seems to be the best possible solution for this problem. A relevant approach which in Chapter 2 namely "automatic qualitative bond graph modelling", proposed by Xia *et. al.* [1991], has already addressed this problem. There, the completeness of a qualitative model is investigated by several physical constraints. How to extend the probing method to provide appropriate modelling detail-level for control aspects will be future work for this hybrid qualitative and quantitative control methodology.

Finally, there is no formal representation in bond graphs to express the pure time-delay of systems. This problem arises because the bond graph modelling process always decomposes a physical system into several standard elements to construct its model. However, pure time delay is usually caused by the aggregation of these elements and the locations of sensors. It is difficult to indicate which element causes how much time-delay to the system. Currently, the control algorithms generated from qualitative bond graph models contain no time-delay information. The coupled tanks liquid level control rig used in this chapter has a time-delay of about two seconds (two sampling periods). The influence of this short time-delay can be compensated by tuning the scaling factors of the controller. Yet, simply tuning the scaling factors is not valid for a control system which has a long time-delay. How to represent the pure time delay in bond graph models is a substantial problem to be resolved.

A topic arising from this control approach is the concept of automatic controller design. As discussed in Chapter 2, the aspect of automatic bond graph modelling has already been developed by some researchers. The general procedure for producing control algorithms from qualitative bond graph models has also been proposed in this and the previous chapter. Combining these two approaches will create an automatic controller design tool. Since this design tool needs no specific parameter details of plants, manpower involved in the controller design process can be much reduced. Thus, a highly automatic system can be achieved, which is very useful for industrial control. Due to the requirements of the fast changing consumer market and operation efficiency, industrial production lines and processes need to be frequently reconfigured, so various control strategies are required. In this circumstance, conventional controllers usually need to be re-designed or re-adjusted by control engineers. On the other hand, making use of automatic controller design tool needs only the information about system structure changes to produce new control strategies. The controller re-adjustment work will be simplified and can be handled by novice or infrequent operators.

CHAPTER 5

AUTO-TUNING

5.1   INTRODUCTION

Chapter 4 has illustrated a hybrid qualitative and quantitative control methodology based on qualitative bond graph reasoning. The controllers constructed through the qualitative controller design procedure possess a basic capability to regulate general feedback control systems. In this chapter, an auto-tuning scheme is developed to enhance the dynamic control performance for the hybrid controllers. One purpose of this auto-tuning scheme is to adjust the controllers automatically to satisfy the performance requirements; the other purpose is to modify the controllers' behaviour automatically in response to changes in the dynamics of the systems and disturbances.

The conventional control design starts with understanding a controlled system and translating dynamic performance requirements into time, frequency, or pole-zero specifications [Franklin *et al.*, 1986]. Understanding a system, what the system is intended to do, how much system error is permissible, what the physical capabilities and limitations are, and how to describe the class of command and disturbance signals to be expected, are the elements in controller design for building individual controllers in various applications. The representation of "understanding" is a specification that the system has a step response inside some constraint boundaries in time domain, an open-loop frequency response satisfying constraints in the frequency domain, or closed-loop poles to the left of some constraint boundary on an *s*-plane. Thus, a controller can be designed according to the specified requirements. The control performance is then verified by applying the design to regulate the actual system or its accurate physical model.

As seen in Chapter 4, the design method of hybrid qualitative and quantitative controllers is different from the conventional one. The control algorithms of a hybrid controller are designed based on the system structure without considering the dynamic performance requirements. The reason is that the control algorithms are derived directly from a qualitative model of the controlled system and are used to represent the control strategy qualitatively, there is no place for a numerical performance specification to conduct the generation of the qualitative control algorithms. Therefore, an original hybrid controller may not fulfil the performance requirements, and hence needs to be tuned to achieve the required specification. In Chapter 4, the design performance is verified through tuning controllers' scaling factors design experience. Hence, the first attempt in this chapter is to develop a general auto-tuning method to automate the verification for the hybrid controllers to complete the automatic controller design process. The tuning function will keep on working during process control so that the controller can adapt to the possible changes in system dynamics, and this is the second attempt in this chapter.

In contrast to most parameter adaptive control algorithms [Åström and Wittenmark, 1989], the auto-tuning scheme developed in this chapter works as a performance adaptive controller and does not require system models and parameter estimation. Performance requirements are given by the performance indexes based on control actions and system responses. Then, the system behaviour is monitored via comparing the observed values to the given values of these performance indexes in real-time. The tuning is enabled according to the monitored result, and guided by the observed values of the performance indexes to adjust the scaling factors. Thus, the considerations of system modelling, parameter evaluation, and nonlinearities can be excluded from the auto-tuning process, and this auto-tuning method will not be flawed by the complexities of systems. In this respect, this auto-tuning approach is motivated through its simplicity of implementation and reduced controller design complexity. This chapter will combine this auto-tuning scheme with the hybrid controllers to regulate the coupled tanks apparatus and compare the control performance to that performed without using the auto-tuning method.

5.2   PERFORMANCE MONITORING

Fig. 5.1 shows the configuration of the auto-tuning hybrid qualitative and quantitative controller. This controller is composed of two loops — inner and outer. The inner loop consists of the controlled system and an ordinary feedback controller, while the outer loop contains an adjustment mechanism and works to adjust the scaling factors of the feedback controller. Here, the adjustment mechanism has two major functions: (1) performance monitoring and (2) scaling factors adjustment. This section will firstly propose the method of performance monitoring, while the scaling factors adjustment will be discussed in the next section.

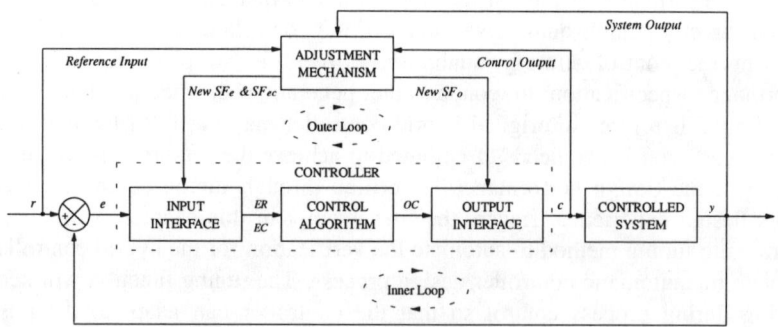

Fig. 5.1 Configuration of Auto-tuning Hybrid Qualitative and Quantitative Controller

In this auto-tuning scheme, control performance is specified by the system output and control actions. In the transient state, the system output is required to track input changes correctly and have a fast response without big overshoot; while in the steady state, the criteria for the system output is small offset error. Moreover, as discussed in Chapter 4, the controller output should be as smooth as possible in both the transient and steady states. The performance indexes of integral of error (J_e) and integral of squared error (J_{se}) are used here as the performance criteria to monitor the overshoot and the steady-state behaviour of the system output. Since this auto-tuning method is applied to digital control systems, the performance indexes are represented as follows:

$$J_e = \sum_{n=1}^{m} \hat{e}(nT), \tag{5-1}$$

$$J_{se} = \sum_{n=1}^{m} \hat{e}^2(nT), \tag{5-2}$$

where $\hat{e}$ denotes the system error which has been obtained by an A/D converter. It can be seen from Eq. (5-1) that a bigger steady-state error will increase the value of

J_e faster (and vice versa), so J_e can be used as the scale of steady-state error. However, in the respect of oscillatory response, positive and negative values of the error will be cancelled by each other and this cancellation will result in a small value for J_e. Therefore, J_e cannot distinguish the difference between a small steady-state error and an oscillatory response. Thus, the performance index J_{se} is employed to detect the system oscillation, since both positive and negative values of the error will increase the value of J_{se}.

Moreover, the transient response is monitored by the average change rate of system error ($J_{\Delta e}$), where

$$J_{\Delta e} = \frac{\sum\limits_{n=1}^{m} \hat{e}c(nT)}{m}.$$ (5-3)

The smoothness of controller output is inspected by the change rate of controller output (J_O) which is given by

$$J_O = \frac{CO(nT)/CO((n-1)T)}{CO((n-1)T)/CO((n-2)T)} = \frac{CO(nT) \cdot CO((n-2)T)}{CO^2((n-1)T)}.$$ (5-4)

The values of J_e, $J_{\Delta e}$ and J_O are required to fall inside their upper and lower constraint boundaries, and the value of J_{se} is required to be smaller than its upper boundary. All the constraint boundaries are determined by controller design according to individual application needs. When any one of the indexes exceeds its constraint boundaries the adjustment mechanism will enable the tuning regime to adjust the controller to drive the abnormal behaviour into the acceptable region.

The reference input of the system is taken as an auxiliary measurement of the adjustment mechanism so that its changes can be identified. Hence, the adjustment mechanism has different tuning strategies to adapt to reference input changes and system parameter (or environment) changes.

Furthermore, stability is also a key requirement for a control system. Conventional stability proofs of adaptive systems are based on mathematical analysis and require numerical models of the systems, i.e. Lyapunov stability analysis [Åström and Wittenmark, 1989]. However, numerical models are not considered in the design process of the hybrid controllers, so conventional stability theories cannot be applied to investigate the stability for the qualitative control systems. It is very difficult (or almost impossible) to analyse the system stability via qualitative information. Therefore, stability monitoring is employed in this auto-tuning scheme to observe system performance and investigate the system stability on-line. Here, the performance index J_{se} has the function of stability monitoring. A big value of J_{se} implies that the system tends to be unstable, thus the adjustment mechanism will tune the controller scaling factors to rectify this circumstance. Making use of stability monitoring will guarantee that the system is working stably.

5.3   SCALING FACTOR ADJUSTMENT

5.3.1   The Basic Idea

As discussed in Ex. 4.4, the hybrid qualitative and quantitative controller can be seen as an enhanced PI controller, and the scaling factors of the hybrid controllers have properties similar to the parameters of PI controllers. Therefore, the tuning strategy is developed to imitate the parameter determining process of PI controllers to adjust the scaling factors for the hybrid controllers.

Let us now consider a conventional PI controller:

$$c(t) = K_p\left(e(t) + \frac{1}{T_i}\int_0^t e(t)dt\right),$$

(5-5)

where c is the control variable, e is the system error, K_p is the feedback gain, and T_i is called the reset time and $1/T_i$ is referred to as the reset rate [Franklin *et al.*, 1986]. Increasing the value of K_p increases the magnitude of the control signal, which is instrumental in reducing the system error and decreasing the rise time of the system. In other words, bigger values of K_p makes the controller output more sensitive to the system error. However, a decreasing damping ratio goes along with the faster response. For higher-order systems, very large values of K_p will often lead to instability. On the other hand, the major effect of the integral control is that it increases the order of the system by one, which will reduce or eliminate the steady-state error of the system. However, this benefit typically comes at the cost of reduced stability. In general, systems will become less stable or less damped when the gain K_p/T_i of the controller is increased.

Corresponding to the PI controller, the hybrid qualitative and quantitative controller also contains the components of proportional amplifiers and integrators. The error change terms in the control algorithm can be seen as the proportional part while the error terms are regarded as the integral one. Thus, it can be seen from the discussion in Section 4.4 that the inverse of output scaling factor $1/SF_o$ in the hybrid controller, which manipulates the magnitude of the system, has the same effect as the feedback gain K_p in the PI controller, and the inverse of error scaling factor $1/SF_e$ plays the same role as the reset rate $1/T_i$. The inverse of error change scaling factor $1/SF_{ec}$ has no corresponding parameter in the PI controller, but it can also be seen as a constant gain of the proportional part for a qualitative control algorithm.

The tuning strategy can be derived according to the characteristics of the scaling factors. In the transient state, a value of $J_{\Delta e}$ which is smaller than its lower constraint boundary means that system response is too slow. Thus, increasing the feedback gain $1/SF_o$ will make the system have a faster response. In contrast, a value of $J_{\Delta e}$ which is larger than its upper constraint boundary indicates that the feedback gain is excessive. An excessive feedback gain is disadvantageous for the stability, so its value should be decreased. Furthermore, the value of J_o concerns the smoothness of controller output and implies the controller's sensitivity to system error changes. If its value exceeds the constraint boundaries, then the proportional constant $1/SF_{ec}$ should be decreased to reduce the oscillation of the controller output. In the steady state, a large value of J_e indicates a large steady-state error. Thus, the integral constant $1/SF_e$ should be increased to reduce the steady-state error. However, if a large value of J_{se} appears together with a small J_e, then it means that the system has a serious oscillatory response. Thus, the integral constant $1/SF_e$ should be decreased to improve the stability for the system.

Several auto-tuning methods have been developed for setting PI and PID controllers and are widely used in the process control industry, i.e. Ziegler-Nichols method [Ziegler and Nichols, 1942], relay feedback method [Åström and Wittenmark, 1989], and the pattern recognition method developed in Foxboro Co. [Kraus and Myron, 1984]. A common property of these tuning methods is that their developments are motivated to optimally setting the PID parameters rather than based on practical performance requirements. Therefore, these methods may not perform a satisfactory tuning for some special applications; for example, a system requiring very fast step response without regard to the overshoot. For this reason, the determination of controller parameters in our tuning mechanism is based on fulfilling the performance requirements and is guided by the ratios of performance index observations to designed performance index specifications. Thus, this tuning method will result in a satisfactory performance rather than an optimum one.

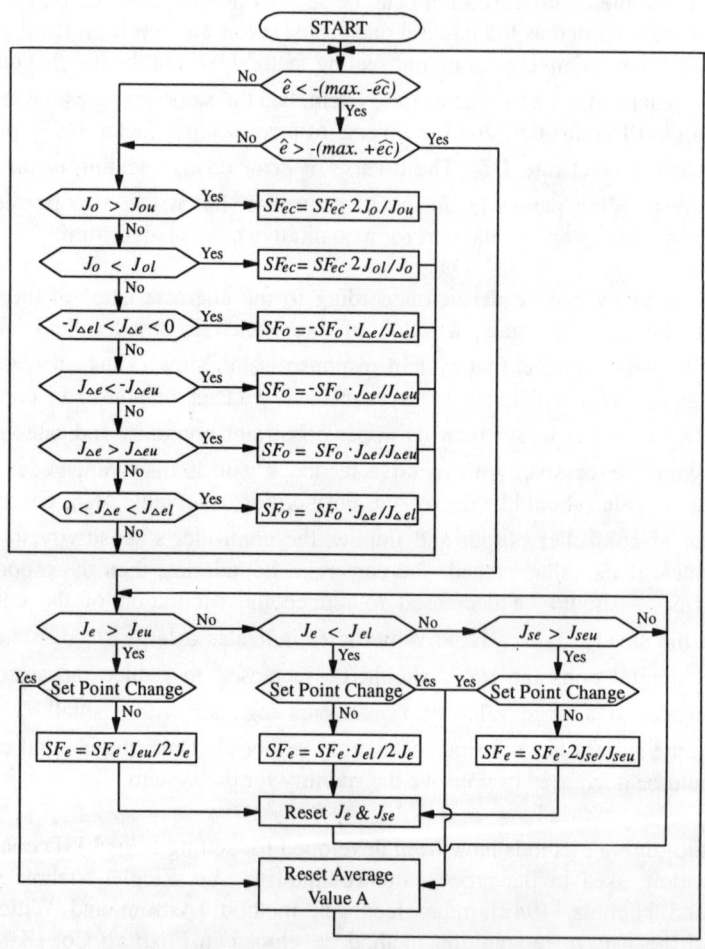

Fig. 5.2 Flowchart of Adjustment Mechanism

5.3.2   Tuning Algorithm

Fig. 5.2 illustrates the flowchart of the adjustment mechanism, where J_{eu}, J_{seu}, $J_{\Delta eu}$, and J_{ou} represent respectively the upper limits of J_e, J_{se}, $J_{\Delta e}$, and J_o, and J_{el}, $J_{\Delta el}$, and J_{ol} denote their lower boundaries respectively.

This adjustment process is composed of two stages: one is tuning the feedback gain $1/SF_O$ and the proportional constant $1/SF_{ec}$, the other is tuning the reset rate $1/SF_e$. The decision whether to tune the $1/SF_O$, $1/SF_{ec}$, or $1/SF_e$ depends on the system states. In general, $1/SF_O$, and $1/SF_{ec}$ are tuned in the transient state, while $1/SF_e$ is tuned in the steady state. Fig. 5.3, which shows a typical step response with a set-point change, is used to help explain the adjustment process.

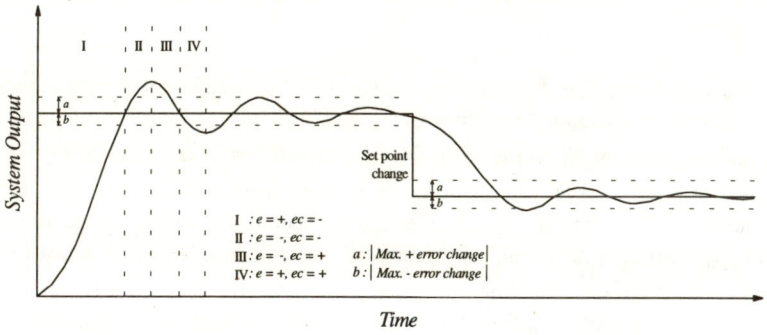

Fig. 5.3 Typical Step Response with the Set-point Change

It is assumed that a control system starts from the origin in Fig. 5.3. The performance monitor firstly investigates the smoothness of system output by examining the value of J_O. If J_O is detected exceeding its constraint boundaries, the adjustment mechanism will increases the value of SF_{ec} by the equation $SF_{ec} = SF_{ec} \cdot 2J_O / J_{ou}$ (when $J_O > J_{ou}$) or $SF_{ec} = SF_{ec} \cdot 2J_{ol} / J_O$ (when $J_O < J_{ol}$). In contrast, if the value of J_O falls within the acceptable region, the performance monitor will then investigate the response speed of the system by inspecting the value of $J_{\Delta e}$. If $J_{\Delta e}$ does not fulfil the requirement, then the value of SF_O will be tuned by the equation $SF_O = -SF_O \cdot J_{\Delta e} / J_{\Delta el}$ (when $J_{\Delta e} < J_{\Delta el}$) or $SF_O = -SF_O \cdot J_{\Delta e} / J_{\Delta eu}$ (when $J_{\Delta e} > J_{\Delta eu}$). Here, the value of $J_{\Delta e}$ is negative when system behaviour is in the zone I but the constraint boundaries of $J_{\Delta eu}$ and $J_{\Delta el}$ are given positive, so these adjustment values are multiplied by a negative sign to keep SF_O

positive. When the system reaches the state where the system error is smaller than the absolute value of maximum negative error change, the adjustment mechanism will switch to monitor the performance index J_e and J_{se}. If the system is lightly damped and gives oscillatory response which is not acceptable, the tuning mechanism will increase the value of SF_e to of $SF_e = SF_e \cdot 2\, J_{se} \,/\, J_{seu}$ to reduce the integral constant of the controller. On the other hand, if the offset error exceeds its constraint boundaries, then the integral constant will be increased by the equation $SF_e = SF_e \cdot J_{eu} \,/\, 2 J_e$ (when the offset error is positive) or $SF_e = SF_e \cdot J_{el} \,/\, 2 \cdot J_e$ (when the offset error is negative) to reduce the error. After the scaling factor SF_e has been tuned, the values of J_e and J_{se} must be reset to zero, otherwise their large values which have already exceeded the constraint boundaries will trigger the tuning of SF_e every sampling period. The coefficients of these tuning equations are obtained from experiments via trial-and-error. Therefore, these coefficients are suitable for the coupled tanks liquid level control rig used in this book, but may be not suitable for other systems.

Once the system reaches the state $\hat{e} < -(max. - \hat{e}c)$, the adjustment mechanism will neither turn to investigate the values of J_o and $J_{\Delta e}$, nor to tune the scaling factors SF_{ec} and SF_o except when the set-point is changed. As in the case of Fig. 5.3, the changed set-point causes a negative error whose absolute value is bigger than the maximum positive error change, so the tuning mechanism will again go to monitor J_o and $J_{\Delta e}$ till the system reaches the condition $\hat{e} > -(max. + \hat{e}c)$. When a set-point change is detected, the adjustment mechanism will firstly reset the average value of past controller outputs A (Eq. 4-17). In the hybrid qualitative and quantitative controller, the value A is added to the controller output change to avoid steady-state error, and to reduces the activity of the controller output. Also, adding A to the controller output change represses the adaptation to set-point changes. The value A needs to be changed to suit the new set-point. The new value of A is obtained according to the new set-point via the method used to evaluate the initial value of A, proposed in Section 4.4.3.

5.4   CASE STUDY

The implementation of the auto-tuning scheme is illustrated by the same SISO and MIMO coupled tanks liquid level control rigs as used in Chapter 4. The schematic diagrams of the liquid level control rigs have been shown respectively in Figs. 4.5 and 4.22, and their detailed specifications have also been described in Sections 4.5.1 and 4.5.2. The experiments illustrated in this section are employed to test three major features of the auto-tuning scheme: (1) the ability of adjusting controllers to fulfil performance requirements, (2) the adaptability to set-point changes, and (3) the adaptability to system parameter changes.

5.4.1 · SISO Cases

Ex. 5.1: *Adjusting Controllers to Fulfil Performance Requirements*

The theme of these experiments is to let the controller tune its scaling factors automatically to regulate Tank 2 liquid level to satisfy the performance requirements. In the following experiments, the discharge tap of Tank 2 (R4) was half opened, and the performance criteria were given by $J_{ou} = 1.04$, $J_{ol} = 0.96$, $J_{\Delta eu} = 10.5$, $J_{\Delta el} = 7.0$, $J_{eu} = +500$, $J_{el} = -500$, and $J_{seu} = 25000$. These performance criteria were obtained from the computations as follows. In the transient state, the oscillation of the controller output was limited to be smaller than 2%. Let $CO((n-1)T) = 1$, then $J_{ou} = 1.02 \cdot 1.02/1 \approx 1.04$, and $J_{ol} = 0.98 \cdot 0.98/1 \approx 0.96$. On the other hand, it was found that if the average error change rate is bigger than 1 mm/sec, the overshot will be bigger than the limitation of steady-state error (2.5%). Thus, $J_{\Delta eu} = (1mm/200mm) \cdot 2048$ (digital input range) ≈ 10.5. The lower limit of the rising slope of the system output ($J_{\Delta el}$) was given be 7.0 directly to obtain an acceptable rise time. In the steady state, the steady-state error was limited to be smaller than 2.5%. Thus, $\pm 2.5\% \cdot 2048 \approx \pm 50$. Further, the values of the performance indexes were reset every ten sampling periods, so $J_{eu} = 50 \cdot 10 = 500$, $J_{el} = -500 \cdot 10 = -500$, and $J_{seu} = 50^2 \cdot 10 = 25000$.

Firstly, the controller began to regulate the system with the values of SF_e, SF_{ec}, and SF_o all given by 1. After ten seconds from the beginning, the performance monitor was initiated to investigate the system behaviour and to introduce the adjustments of scaling factors. Two control and adjustment results for different set-points are shown in Figs. 5.4. and 5.5. The tuning result of the experiment illustrated in Fig. 5.4 was $SF_e = 1.0$, $SF_{ec} = 49.92$, and $SF_o = 0.58$, while the result of the experiment illustrated in Fig. 5.5 was $SF_e = 1.0$, $SF_{ec} = 50.85$, and $SF_o = 0.53$. It can be seen that these adjustment results are quite similar. That is, the setting for scaling factors is related to the performance requirements rather than the system set-points.

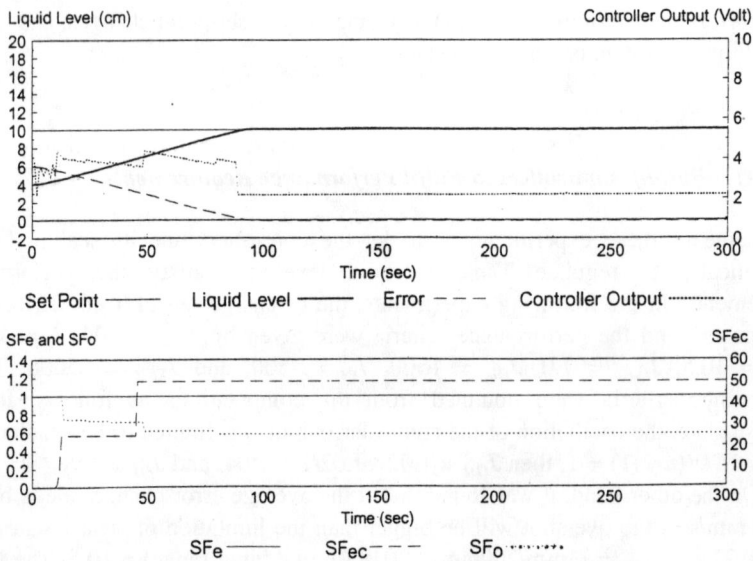

Fig. 5.4 Step Response of Tank 2 Liquid Level and Scaling Factor
Adjustments;Set-point = 10 cm.

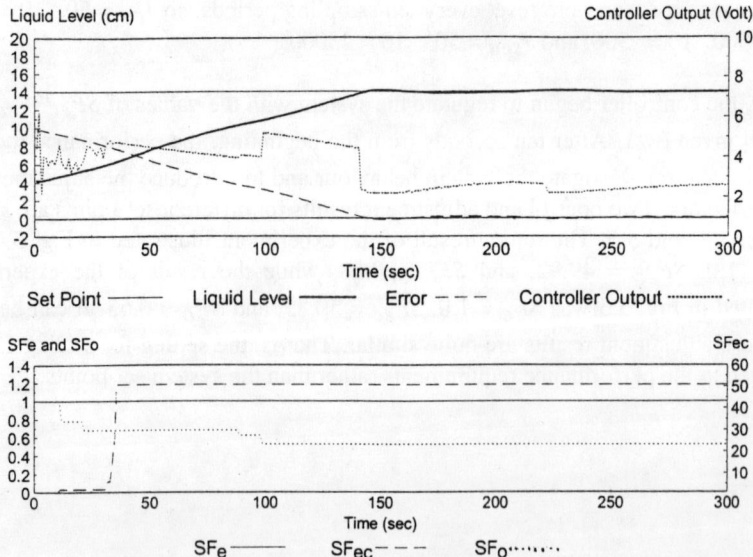

Fig. 5.5 Set Response of Tank 2 Liquid Level and Scaling Factor
Adjustments; Set-point = 14 cm.

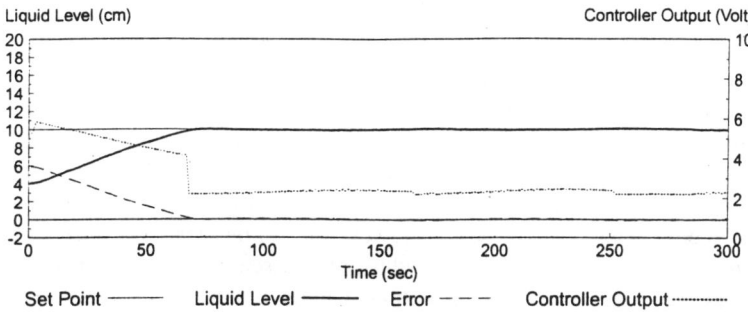

Fig. 5.6 Step Response of Tank 2 Liquid Level;
Set-point = 10 cm, $SF_e = 1$, $SF_{ec} = 49.92$, $SF_0 = 0.58$, $\delta = 0.9$.

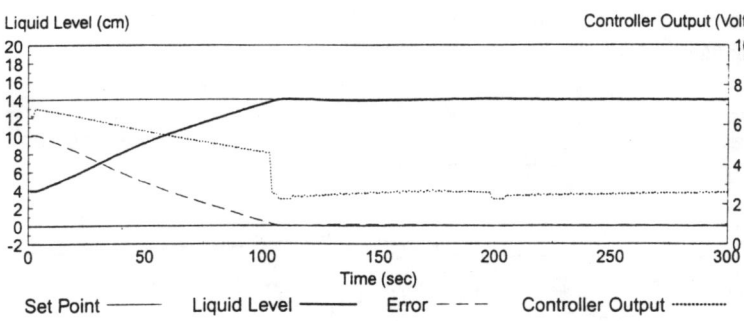

Fig. 5.7 Step Response of Tank 2 Liquid Level;
Set-point = 14 cm, $SF_e = 1$, $SF_{ec} = 50.85$, $SF_0 = 0.53$, $\delta = 0.9$.

Figs. 5.6 and 5.7 show the control results of using the final tunings obtained from the previous experiments. Here, system behaviours can meet the performance requirements without any re-tuning. That is, the adjustment mechanism can complete effectively the tuning task in one test run. Although the control performance tuned by the auto-tuning scheme is not as good as that tuned by the human operator (shown in Fig. 4.9), it has resulted in a satisfactory performance which can fulfil the performance criteria.

Ex. 5.2: *Adaptability to Set-point Changes*

The following experiments were used to test whether the auto-tuning scheme can improve the adaptability to system set-point changes. The set-point of Tank 2 liquid level was changed from 10 cm to 11 cm at 250 sec and then drawn back to 10 cm at 450 sec. The discharge tap of Tank 2 was half opened.

Fig. 5.8 shows the control result without using the auto-tuning scheme, while Fig. 5.9 shows the control result obtained with the use of auto-tuning function. The performance criteria used here were the same as that in Ex. 5.1. By comparing these results, it can be found that the controller which employed the auto-tuning scheme can track the set-point changes better. However, it can be seen from Fig. 5.9 that the better adaptability is achieved not by adjusting the scaling factors but by resetting the average value of past controller outputs (A) to its initial value $A(0)$. There are two reasons for this tuning approach: one is that adding A to the controller output change has an effect on keeping the controller output stable, so resetting its value will make the controller track the set-point changes more effectively; the other is, as discussed in Ex. 5.1, that the settings of scaling factors do not depend on the values of set-points, so the adjustment mechanism does not tune the scaling factors for set-point changes.

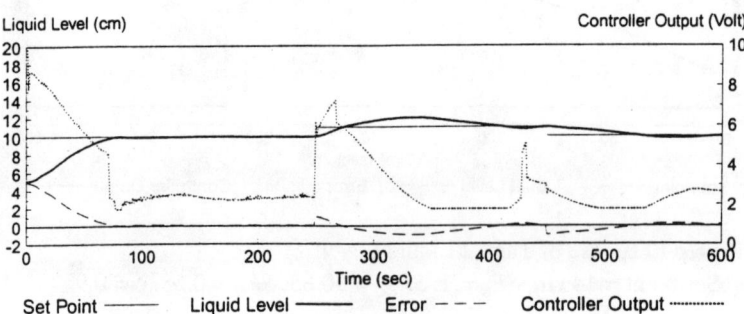

Fig. 5.8 Step Response of Tank 2 Liquid Level with Set-point Changes, and without Using the Auto-tuning Scheme; $SF_e = 0.5$, $SF_{ec} = 25$, $SF_o = 0.5$, $\delta = 0.9$.

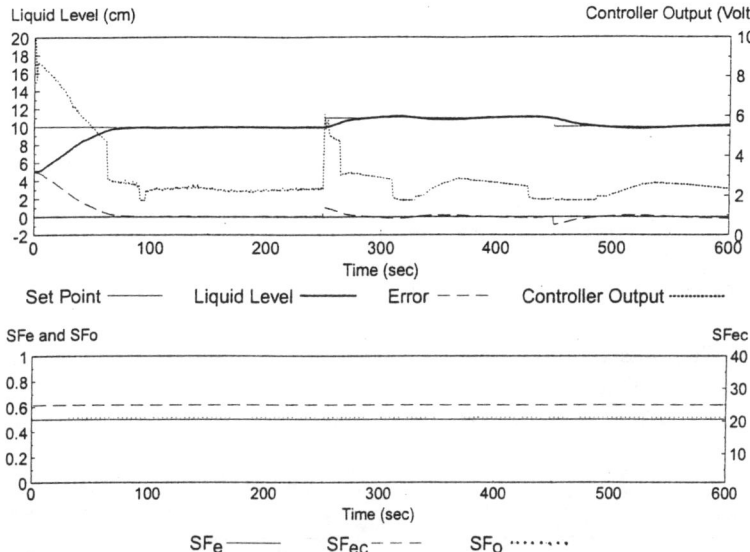

Fig. 5.9 Set Response of Tank 2 Liquid Level and Scaling Factor Adjustments with Set-point Changes

Ex. 5.3: *Adaptability to System Parameter Changes*

The following experiments were used to test the adaptability of the auto-tuning scheme to system parameter changes. These experiments began with the discharge tap of Tank 2 (*R4*) half opened and then the parameter change was made by the tap being fully opened at 250 sec. Figs. 5.10 and 5.11 show respectively the system behaviours regulated by the controllers without and with the auto-tuning function. It can be seen from Fig. 5.10 that the hybrid qualitative and quantitative controller already has good robustness to system parameter changes, so the auto-tuning scheme provided only a slightly better control performance in terms of steady-state error. However, the auto-tuning scheme still demonstrated an adequate tuning for the system parameter change. As shown in Fig. 5.11, the value of the output scaling factor was decreased to increase the feedback gain for the controller to cope with the increasing outlet flow quantity.

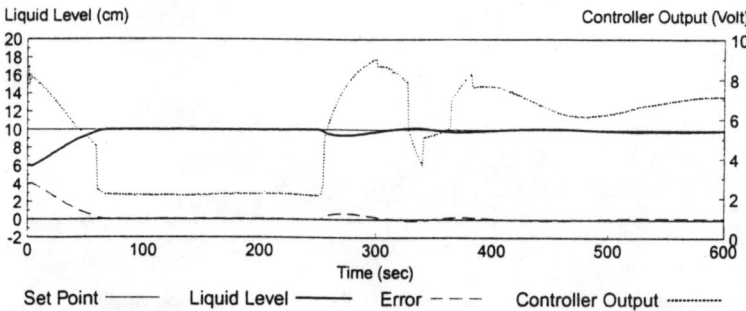

Fig. 5.10 Step Response of Tank 2 Liquid Level with System Parameter Change,
and without Using the Auto-tuning Scheme;
$SF_e = 0.5$, $SF_{ec} = 25$, $SF_o = 0.5$, $\delta = 0.9$.

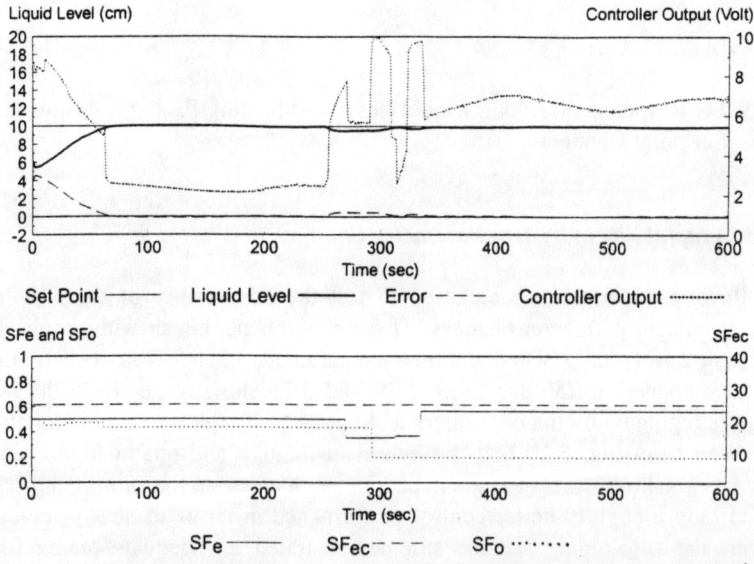

Fig. 5.11 Set Response of Tank 2 Liquid Level and Scaling Factor Adjustments
with System Parameter Changes

Fig. 5.12 shows the control result of the experiment beginning with the discharge tap
of Tank 2 fully opened, and using the final tuning results of the experiment
illustrated in Fig. 5.11 ($SF_e = 0.56$, $SF_{ec} = 25$, and $SF_o = 0.20$). It can be seen that

although the controller output has been its maximum value in the transient state, the system rising slope still cannot match the performance requirement $J_{\Delta e} < -7.0$. This is because the capacity of the pump is not big enough to supply sufficient flow quantity to increase the liquid level faster when the discharge tap is fully opened. In this case, the values of the scaling factors used here can be still regarded as an adequate choice for the controller, because a further greater feedback gain will not shorten the rise time for the system but may cause the system unstable.

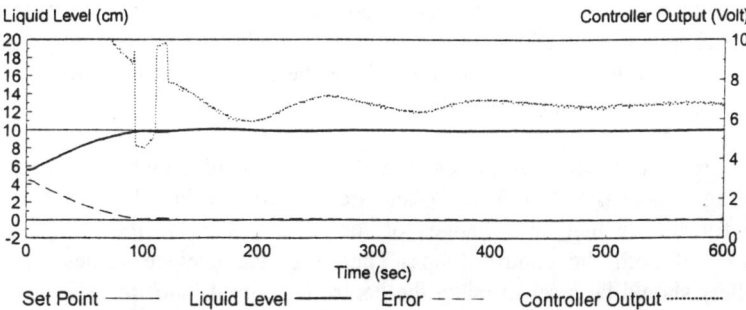

Fig. 5.12 Step Response of Tank 2 Liquid Level with the Outlet Tap Fully Opened; $SF_e = 0.56$, $SF_{ec} = 25$, $SF_o = 0.20$, $\delta = 0.9$.

5.4.2   MIMO Cases

So far the auto-tuning scheme has been successfully applied to the SISO coupled tanks liquid level control rig. In this section, the auto-tuning scheme is extended to regulate the MIMO one. As discussed in Chapter 4, the SISO and MIMO hybrid qualitative and quantitative controllers have the same property — their control algorithms are constructed by the variables of system errors and error changes and can be regarded as enhanced PI controllers. Ideally, the tuning strategy based on PI parameter adjustments (scaling factor adjustments) can also be applied to multivariable adaptive controllers.

The configuration of the MIMO coupled tanks liquid level control rig employed here has been explained in Chapter 4, and the control algorithms of the MIMO system have been stated in Eqs. (4-92) and (4-93). There are four system output variables: error of Tank 1 liquid level (ER_1), error of Tank 2 liquid level(ER_2), error change of Tank 1 liquid level (EC_1) and error change of Tank 2 liquid level (EC_2), and two controller output variables: controller output change of Controller 1 (OC_1) which regulates the flow fed into Tank 1, and controller output change of Controller

2 (OC_2) which regulates the flow fed into Tank 2. All state variables in the control algorithms have their corresponding scaling factors and performance indexes. The MIMO auto-tuning is performed based on monitoring the performance indexes and tuning the corresponding scaling factors when any of the indexes exceed the constraint boundaries. The adjustment mechanism is the same as that used for SISO systems, which has been illustrated in Fig. 5.2. According to the control algorithms Eqs. (4-94) and (4-95), it can be inferred that tuning either of the error scaling factors will affect both the control loops, while tuning the error change or output scaling factors will only affect their corresponding control loop. For example, tuning the error change scaling factor for the variables of Tank 1 liquid level will has an effect on the response of the controller which regulates the pump feeding liquid into Tank 1, but does not affect the other controller.

Besides, in these MIMO cases, after each of the error scaling factors has been tuned, the average values (A) of both controllers are reset to their initial values ($A(0)$). The reason for this is that any changes of the error scaling factors will affect the responses of both the control loops. Therefore, the average values (A) of the controllers should be reset to make the controller outputs shift to their new values faster to adapt to system changes.

Ex. 5.4: *Adjusting Controllers to Fulfil Performance Requirements*

The theme of these experiments is to tune the controllers' scaling factors automatically to regulate both the Tank 1 and Tank 2 liquid levels within performance requirements. In the following experiments, the discharge tap of Tank 2 was fully opened. The performance criteria were given by $J_{ou1} = 1.04$, $J_{ol1} = 0.96$, $J_{ou2} = 1.04$, $J_{ol2} = 0.96$, $J_{\Delta eu1} = 24.0$, $J_{\Delta el1} = 19.0$, $J_{\Delta eu2} = 24.0$, $J_{\Delta el2} = 19.0$, $J_{eu1} = +700$, $J_{el1} = -700$, $J_{eu2} = +700$, $J_{el2} = -700$, $J_{seu1} = 49000$, and $J_{seu2} = 49000$, where the performance indexes with the subscript of 1 were used to monitor the behaviour of Tank 1 and Controller 1, while the ones with the subscript of 2 were employed to monitor the behaviour of Tank 2 and Controller 2. How to evaluate the performance criteria has been explained in Ex. 5.1.

The experiment began with all the scaling factors being set to 1. After ten seconds from the beginning, the adjustment mechanism was enabled to monitor the control performance and adjust the scaling factors. Fig. 5.13 shows the experimental results, where the final tuning of the scaling factors were $SF_{e1} = 1.0$, $SF_{ec1} = 39.55$, $SF_{o1} = 0.68$, $SF_{e2} = 1.0$, $SF_{ec2} = 17.57$, and $SF_{o2} = 1.012$. Next, these tuning results were applied to regulate the coupled tanks system, and the experimental result is shown in Fig. 5.14. It can be seen that the adjustment mechanism can fulfil the control performance requirements for the MIMO system via tuning the scaling factors.

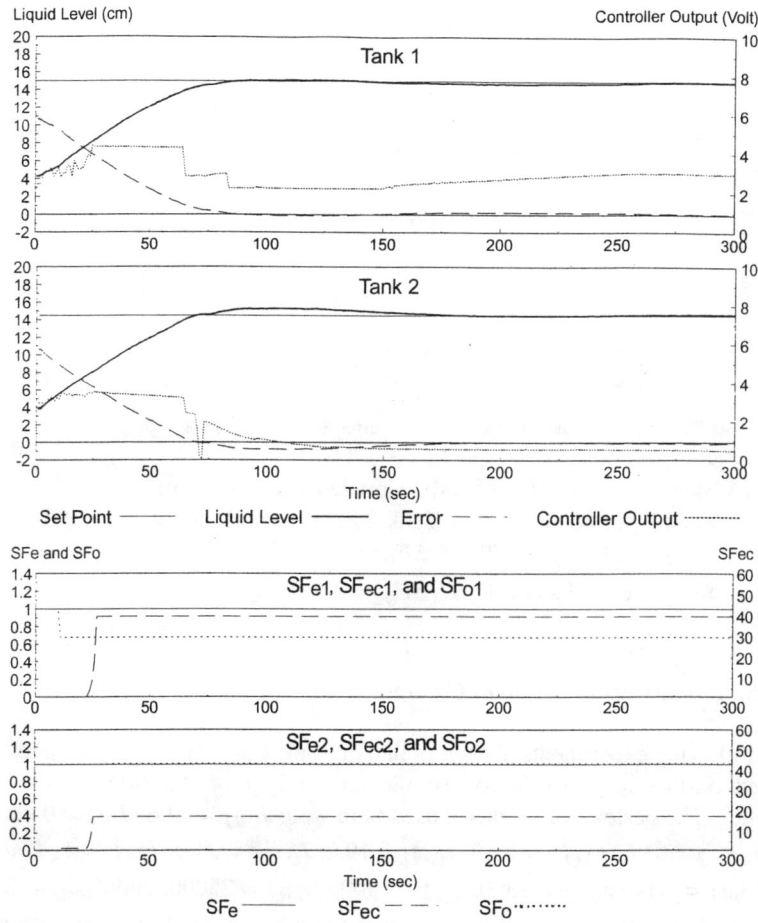

Fig. 5.13 Step Response of the MIMO system and Scaling Factor Adjustments;
Tank 1 Set-point = 15 cm, Tank 2 Set-point = 14.5 cm,

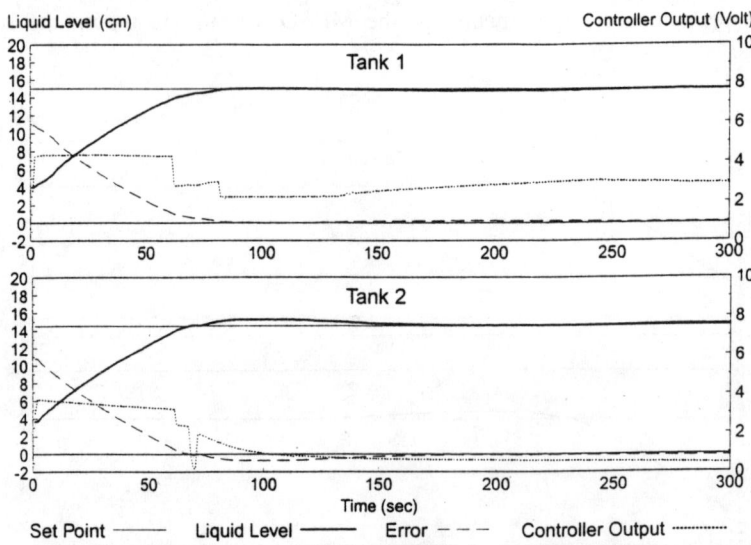

Fig. 5.14 Step Responses of the MIMO Coupled Tanks System;
Tank 1 Set-point = 15 cm, Tank 2 Set-point = 14.5 cm,
$SF_{e1} = 1.0$, $SF_{ec1} = 39.55$, $SF_{o1} = 0.68$, $\delta_1 = 0.9$,
$SF_{e2} = 1.0$, $SF_{ec2} = 17.57$, $SF_{o2} = 1.012$, $\delta_2 = 0.9$.

Ex. 5.5: *Adaptability to Set-point Changes*

In the following experiments, the set-points of Tank 1 and Tank 2 liquid levels were changed on-line respectively to test the adaptability of the MIMO auto-tuning controller. The performance criteria used here were $J_{ou1} = 1.04$, $J_{ol1} = 0.96$, $J_{ou2} = 1.04$, $J_{ol2} = 0.96$, $J_{\Delta eu1} = 24.0$, $J_{\Delta el1} = 19.0$, $J_{\Delta eu2} = 24.0$, $J_{\Delta el2} = 19.0$, $J_{eu1} = +400$, $J_{el1} = -200$, $J_{eu2} = +300$, $J_{el2} = -200$, $J_{seu1} = 25000$, and $J_{seu2} = 23500$. Fig. 5.15 shows the system behaviour responding to set-point changes without the auto-tuning, while Fig. 5.16 shows the system behaviour when the auto-tuning scheme was employed to adapt to the set-point changes. It can be seen by comparing Figs. 5.15 and 5.16 that the auto-tuning scheme improved the control performance effectively for the MIMO controller when set-points were changed.

As discussed in Ex. 4.5, the distance between the liquid levels in Tank 1 and Tank 2 is limited by the capacity of the discharge tap of Tank 2. The maximum liquid level distance (around 1 cm) happens when the pump of Tank 2 stops pumping. In other words, the controllers cannot force the distance between the liquid levels to exceed 1 cm. Therefore, the set-point changes given in these experiments were limited within ±0.5 cm, otherwise the controllers will fail to tack the set-point changes.

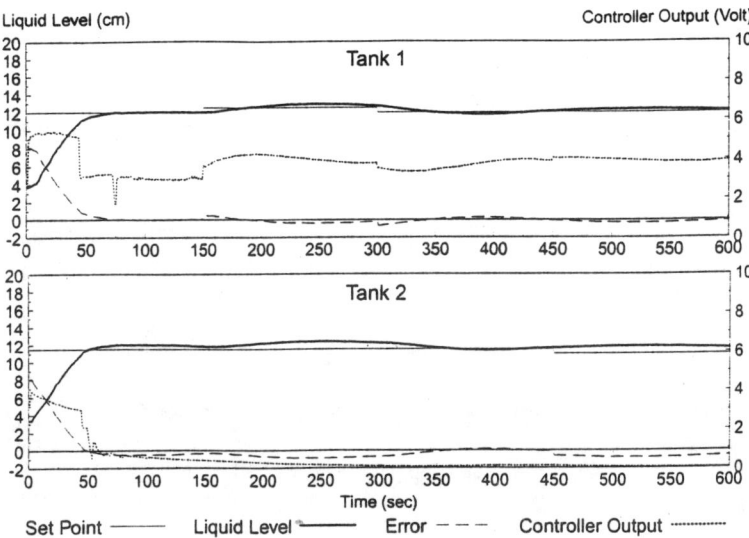

Fig. 5.15   Step Responses of the MIMO Coupled Tanks System
with Set-point Changes and Without Auto-tuning;
$SF_{e1} = 0.5, SF_{ec1} = 30, SF_{o1} = 0.5, \delta_1 = 0.9,$
$SF_{e2} = 0.5, SF_{ec2} = 30, SF_{o2} = 1.0, \delta_2 = 0.9.$

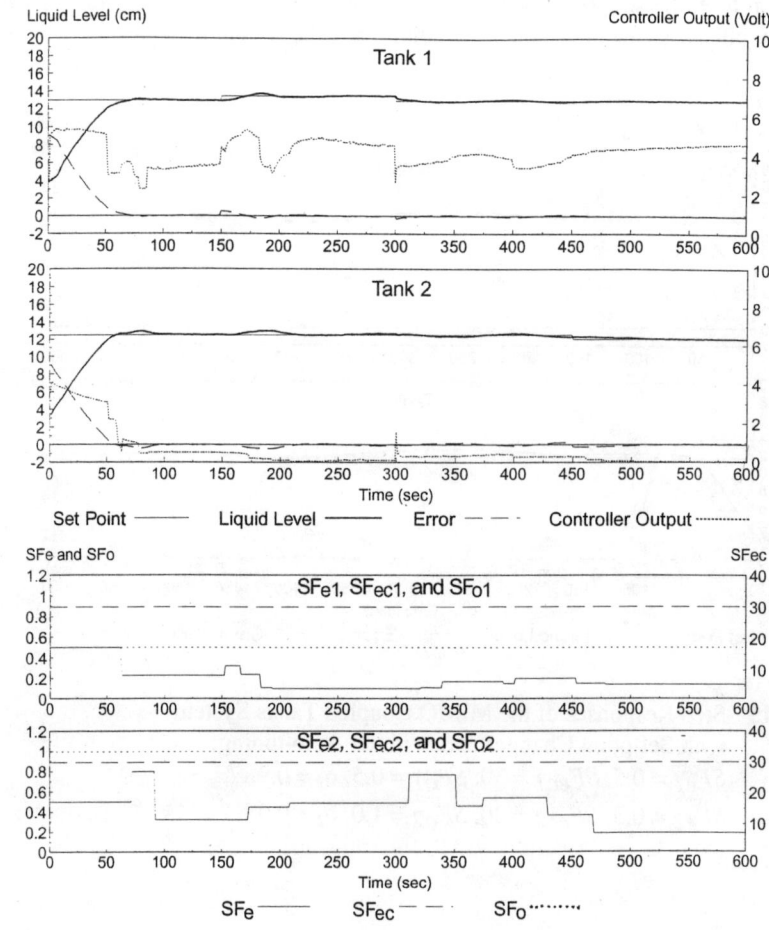

Fig. 5.16 Step Response of the MIMO system and Scaling Factor Adjustments with Set-point Changes and Using the Auto-tuning Function

Ex. 5.6: *Adaptability to System Parameter Changes*

In these experiments, the parameter of the Tank 2 discharge tap was changed to examine the controller's adaptability to system parameter changes. The performance criteria were the same as that in Ex. 5.5. In the first two experiments, the discharge tap was initially fully opened, and then a slight leakage was introduced to Tank 2 at 250 sec, and finally the leakage was repaired at 450 sec. Figs. 5.17 and 5.18 show the control results respectively without and with using the auto-tuning scheme. In the second two experiments, the discharge tap was also initially fully opened, and then was closed gradually at 250 sec, and was fully opened again at 450 sec. Fig. 5.19 shows the system behaviour controlled without the auto-tuning function, while Fig. 5.20 shows the experimental result regulated by the auto-tuning scheme. The experimental results show that the auto-tuning scheme improves the adaptability of the MIMO controllers in coping with system parameter changes.

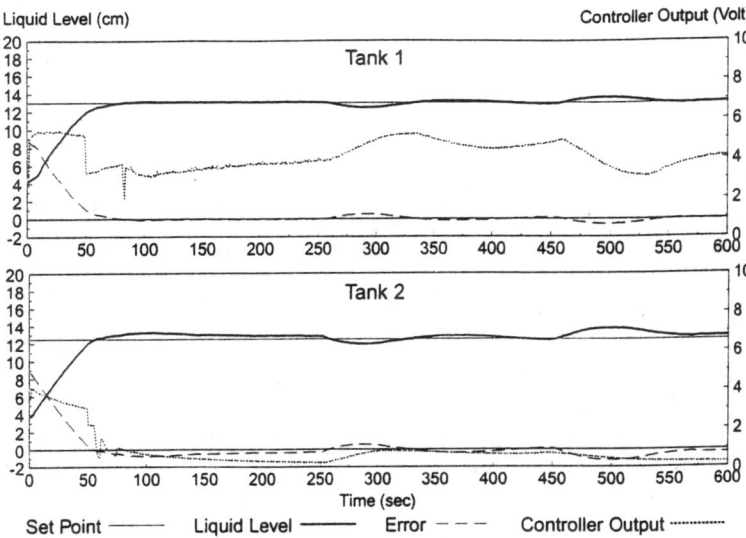

Fig. 5.17   Step Responses of the MIMO Coupled Tanks System without Auto-tuning; Tank 2 leaked at 250 sec and was repaired at 450 sec, $SF_{e1} = 0.5$, $SF_{ec1} = 30$, $SF_{o1} = 0.5$, $\delta_1 = 0.9$, $SF_{e2} = 0.5$, $SF_{ec2} = 30$, $SF_{o2} = 1.0$, $\delta_2 = 0.9$.

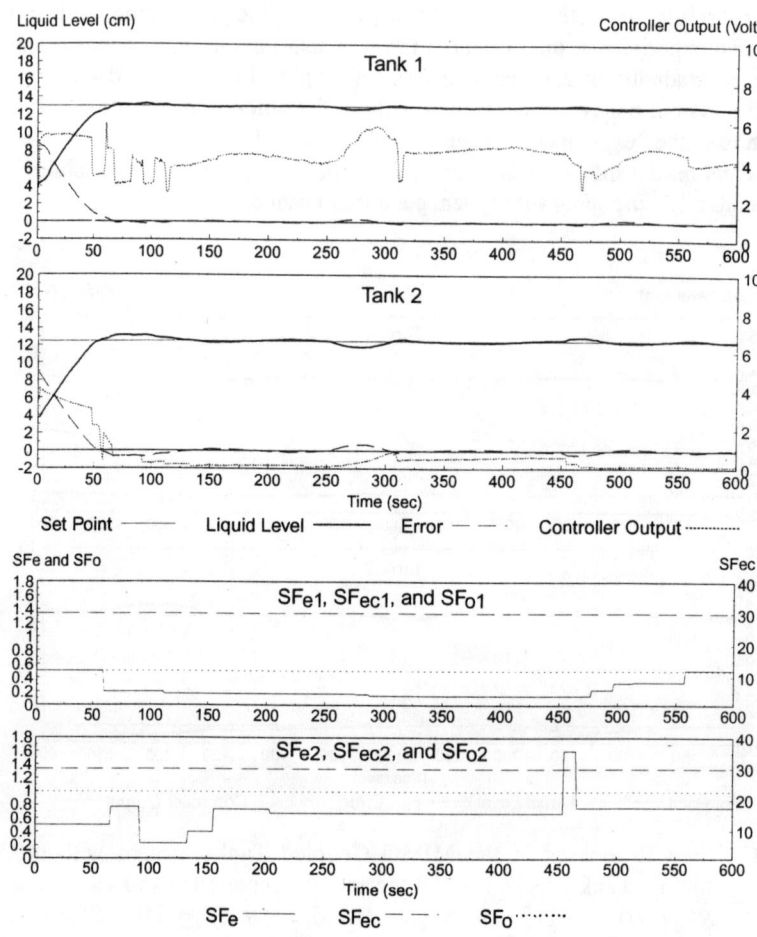

Fig. 5.18 Step Response of the MIMO system and Scaling Factor Adjustments; Tank 2 leaked at 250 sec and was repaired at 450 sec.

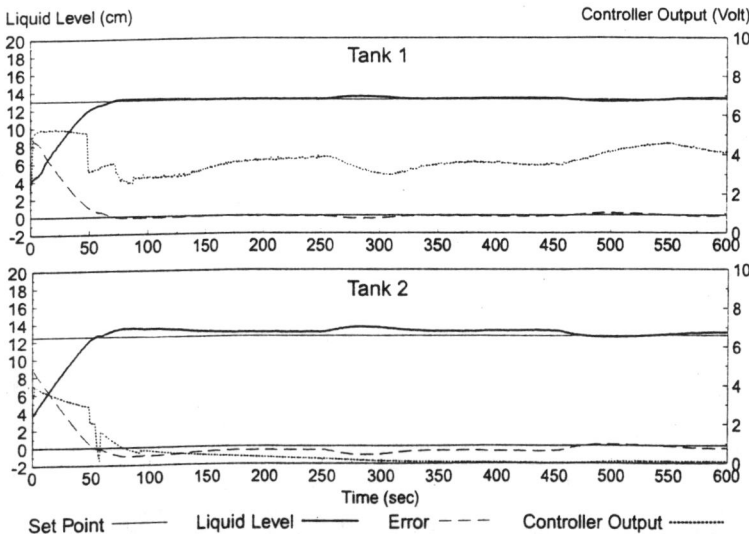

Fig. 5.19 Step Responses of the MIMO Coupled Tanks System without Auto-tuning; Discharge Tap was Closed a Little at 250 sec and fully open at 450 sec,

$SF_{e1} = 0.5$, $SF_{ec1} = 30$, $SF_{o1} = 0.5$, $\delta_1 = 0.9$,

$SF_{e2} = 0.5$, $SF_{ec2} = 30$, $SF_{o2} = 1.0$, $\delta_2 = 0.9$.

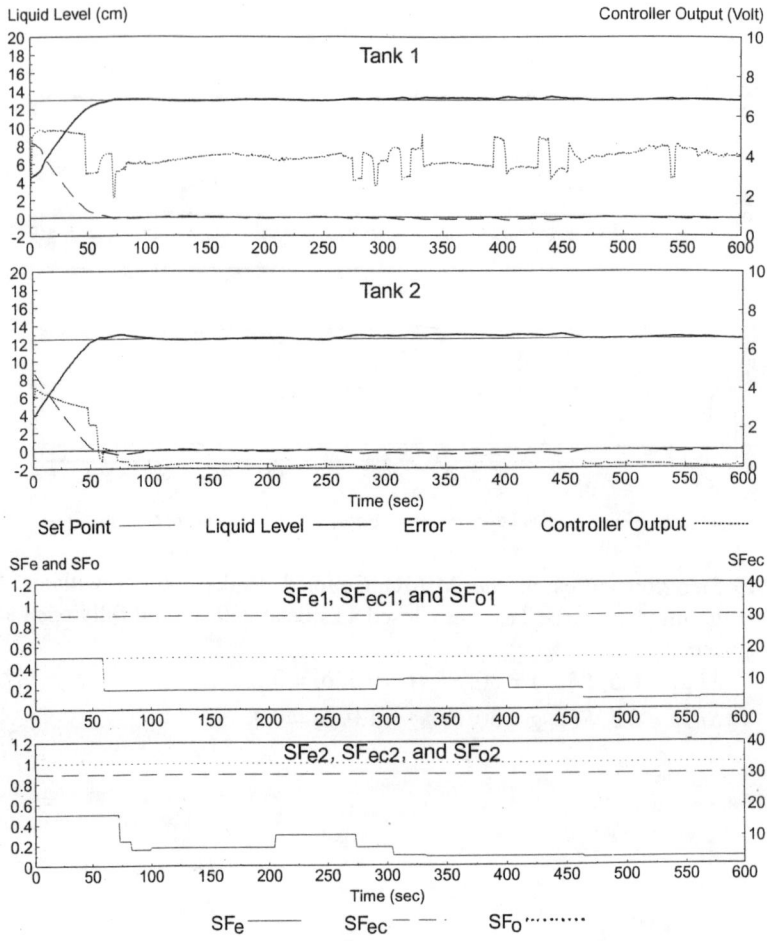

Fig. 5.20 Step Response of the MIMO system and Scaling Factor Adjustments; Discharge Tap was Closed a Little at 250 sec and fully open at 450 sec.

5.5   DISCUSSION

This chapter has presented an auto-tuning scheme developed based on performance adaptation for the hybrid qualitative and quantitative controllers. The integral of error, integral of squared error, change rate of error, and change rate of controller output are used as performance indexes to monitor system performance. Performance requirements of control systems are given in terms of these performance indexes. The tuning of scaling factors is conducted by the ratio of observed values of the performance indexes to the given values of the performance indexes. Some successful experimental results have shown that the auto-tuning scheme can adjust controllers to meet the performance requirements and adapt the changes in system dynamics.

The auto-tuning scheme has two substantial features for the hybrid qualitative and quantitative control method. Firstly, it completes the automatic design process of the hybrid controllers. The systematic controller design procedure proposed in Chapter 4 did not involve the consideration about performance requirements, and the controller adjustments were performed by designers according to their operational experience. This auto-tuning scheme provides a schematic and automatic way to adjust controllers to meet the performance requirements, mainly to keep the controller design procedure simple so that novice or infrequent users can handle it. Secondly, this auto-tuning scheme for the hybrid qualitative and quantitative control method avoids the difficulties caused by system complexity and parameter evaluation when tuning controllers is necessary. As discussed in Chapter 4, the hybrid controllers can regulate systems without considering system parameters, and the controller design process based on qualitative bond graph modelling is practicable for most engineering systems and is not harassed by the complexity of systems. The auto-tuning scheme based on performance adaptation augments these advantages so that it is very suitable to be combined with the hybrid control method.

The key to the success in applying the auto-tuning scheme is the adequate setting of performance requirements. Indulgent requirements, of course, cannot lead to a good performance, while too strict requirements will cause a very active system because of the frequent tuning and may also lead to a poor control performance. Thus, understanding the controlled systems is the very first thing to do for determining the performance criteria. Realising what the system is intended to do and how much system error is permissible will facilitate the primary setting of performance criteria, and understanding what the physical capabilities and limitations are can avoid too strict performance requirements.

Also, there remain some alternatives to enhance the auto-tuning scheme. Firstly, making use of more effective and sensitive performance indexes will more accurately distinguish system dynamic changes so that more fine tuning can be expected. Secondly, since the formulas used for the scaling factor evaluation are given by experience rather than the mathematical relations between the performance indexes and scaling factors, finding such relations will make the tuning more effective. Finally, this auto-scheme and the stochastic adaptive control method have a similar property — they are both based on performance adaptation. Moreover, the structure of the control algorithm of a hybrid qualitative and quantitative controller is also similar to that of a stochastic model [Åström and Wittenmark, 1989]. Therefore, it is possible to employ the existing stochastic adaptive control techniques to further enhance this auto-tuning method.

CHAPTER 6

FAULT DIAGNOSIS

6.1 INTRODUCTION

Fault diagnosis is one of the most important tasks assigned to intelligent supervisory control systems. The diagnosis problem starts with the observation of abnormal behaviour deviating from the expected or desirable conditions. A diagnosis mechanism at this stage is required to explore initial fault candidates which would cause the observed abnormal behaviour, and to suggest measurement selections to help refine the initial candidate set. Most of the early approaches in this field have been rule-based. They used simple production rules to provide a mapping between the possible causes and inputs of a system and the possible faults. Although many implemented rule-based diagnosis systems provided an effective diagnosis performance (e.g. MYCIN), they all suffered from well-known disadvantages like incompleteness and inflexibility. Better approaches to fault diagnosis are based on an underlying model of a device's structure and behaviour. Models provide a knowledge representation about a large amount of structural, functional, and behavioural information and their relationships, which are needed to enable a more complex cause-effect reasoning so that more powerful and robust diagnosis systems can be built.

There are two major types of model-based fault diagnosis: quantitative and qualitative. The quantitative approach requires advanced information processing techniques such as state estimation, parameter estimation, and adaptive filtering. The main problem in the quantitative approach is that it can be applied only when precise numerical models are available; a much more serious problem is that of the sensitivity of the detection system with respect to modelling errors. Logically, the effect of modelling errors obscures the effect of faults and is therefore a source of false alarms [Frank, 1990]. On the other hand, the qualitative approach is more applicable and effective when numerical models are not available. It makes use of a qualitative causal analysis which links individual component malfunctions expressed in a qualitative form with deviations in the measurement values.

An approach to qualitative-model-based fault diagnosis via fuzzy qualitative simulation (FuSim) has been developed by Shen and Leitch [1992] for diagnosing continuous dynamic systems. The diagnosis in this method begins with monitoring the system performance via the synchronous tracking of the system behaviour. A behaviour simulator using FuSim is employed to predict the qualitative behaviour of a system from observations. If the observed behaviour cannot match the predicted one, the candidate generator will then search for the modified models which can derive the behaviour to match the observations. Thus, fault candidates can be found from the modified models. This FuSim-based diagnosis method has several advantages over the methods based on ordinary qualitative simulation (QSim) [Kuipers, 1986]: (1) since FuSim provides more precise information than QSim, this method can derive more accurate diagnostic results; and (2) making use of fuzzy sets allows the subjective element in system modelling to be incorporated and reasoned with in a formal way. However, the difficulties with this method are: (1) how to determine suitable numbers of fuzzy sets for a universe of discourse and appropriate membership functions for these fuzzy sets so that the abnormal behaviour can be distinguished explicitly; and (2) how to choose appropriate modelling dimensions to search for candidate generation and indicate particular interesting fault areas and fault types for bounding the search so that an efficient diagnosis can be achieved.

Another alternative approach to qualitative-model-based fault diagnosis is "reasoning from first principles", which uses a physical model to explain discrepancies between its faulty and normal behaviours. The early work in this area, such as DART [Genesereth, 1984], HT [Davis, 1984], and GDE [de Kleer and Williams, 1987], presented several prototypes for diagnostic systems in the domain of digital circuits. A general theory framework for diagnosis from first principles was proposed by Reiter [1987]. In Reiter's theory, a system is defined as a triple (SD, COMPS, OBS), where SD (the system description) and OBS (the observations) are finite sets of first-order sentences, and COMPS (the components in the system) is a finite set of constants. A diagnosis is a minimal set $\Delta \subseteq$ COMPS such that SD $\cup$ OBS $\cup$ $\{\neg$ AB$(c) \mid c \in$ COMPS $- \Delta\}$ is consistent, where AB is the abnormality predicate, such that AB(c) means that component c is abnormal or faulty. If the OBS shows an abnormal behaviour, then there must be a number of conflict sets (subsets of COMPS) which will cause SD $\cup$ OBS $\cup$ $\{\neg$ AB$(c) \mid c \in$ COMPS $- \Delta\}$ to be inconsistent. Thus, fault candidates can be found from these conflict sets. This theory has also been extended to diagnose dynamic systems, e.g. DIAMON

[Lackinger and Nejdl, 1991] and Inc-Diagnose [Ng, 1990]. These approaches resort to qualitative simulation as an inference engine during the diagnosis process to predict possible behaviour patterns.

The apparent advantages of first principles over rule-based diagnosis approaches are: (1) this method can diagnose faults that have never occurred before, since only knowledge of correct system behaviour is required; (2) this reasoning method employed is domain independent, so it is not necessary to tailor the inference mechanism for different applications; and (3) given a device description, users can design a diagnosis system without associations from human experts. However, there are also several constraints on the first principles method. Firstly, this method can only deal with faulty components. If it is extended to deal with the connections between components, sensors, or actuators, the number of processing times for possible diagnoses will become much too large. Besides, this method is formulated in terms of first-order sentences, so it is necessary to generate a set of complete and consistent first-order sentences for system description. However, representing a complex engineering system with first-order sentences requires a strong background of modelling skills. Therefore, there are relatively few people who can adequately utilise the first principles method.

The second constraint of first principles has been overcome by Biswas *et al* [1992], where a formal qualitative bond graph modelling scheme was employed to generate state equations as the first-order system descriptions for fault diagnosis based on first principles. This chapter adopts the qualitative bond graph modelling scheme proposed in Chapter 3 to resolve the above system representation problem, and develops a new qualitative fault diagnosis method. With the use of the qualitative bond graph modelling method, several advantages of this fault diagnosis methodology are evident. Firstly, the qualitative equations used for system description are easily generated in a systematic way without concern about the causality assignment and algebraic manipulations. Secondly, the qualitative equations explicitly describe the structural information about component locations and their interconnections. Therefore, a diagnosis mechanism can easily localise system faults through reasoning with the relations between the observed behaviour and system structure. Finally, since the qualitative models contain no numerical information, the impairment of modelling errors (a fundamental problem in quantitative model-based fault diagnosis methods) can be eliminated.

The description in this chapter for fault diagnosis includes the following tasks:

- Define a set of mathematical qualitative descriptors to represent system behaviour and component states, and a set of qualitative operators to compute these descriptors. Thus, the qualitative equations used to represent qualitative bond graph models can be utilised as a heuristic algorithm to generate fault candidates.

- Develop an inference mechanism for fault diagnosis based on the models used to generate the control algorithms for hybrid qualitative and quantitative controllers. Therefore, the same model can be applied for both aspects of automatic control and fault diagnosis. Meanwhile, the inference mechanism should be available to diagnose system faults without fault models, or fault trees, so that this diagnosis method can localise the unanticipated faults. In addition, the method should be able to localise all the failures which may occur in either sensors, controllers, power drivers, or components via a simple fault detection method.

- Develop a generic inference methodology to suggest additional measurements to further refine the diagnosis results according to the observation and the structural information provided in qualitative bond graph models.

The application of this qualitative fault diagnosis method is again illustrated by the MIMO coupled tanks liquid level control rig used in the previous chapters, operating in real-time using Modula-2 software code.

6.2   CONFIGURATION OF THE FAULT DIAGNOSIS METHOD

Fig. 6.1 shows the block diagram for the qualitative-model-based fault diagnosis method. The modelling process here is the same as that in the hybrid qualitative and quantitative control method, but the modelling result, a set of qualitative equations, need not be simplified. The reason is that the unsimplified qualitative equations provide the system structural information which is necessary for qualitative fault diagnosis. The qualitative equations have a formulation where the relations between the whole system's behaviour and components' behaviour can be analysed. Therefore, fault candidates can be found by reasoning with the relations between the observed abnormal system behaviour and the assumed faulty component behaviour.

The fault diagnosis mechanism has two major functions: fault detection and fault localisation. The fault detection part contains a performance monitor which employs the performance indexes, J_e and J_{se}, used in the auto-tuning scheme to monitor system behaviour. The observations needed for this method include system errors and error changes, auxiliary measurements, controller outputs, and reference inputs. The auxiliary measurements measure some internal states which are not adopted as feedback signals, and these measurements are very helpful in reducing the number of fault candidates. The controller outputs are employed to help find controller faults, while reference inputs are used to distinguish whether a rapid change in the system behaviour is caused by reference input changes or by system faults. When the values of J_e and/or J_{se} exceed their constraint boundaries, it can be presumed that some faults have occurred in the system. Then, the observations will be converted to their corresponding qualitative values so that the qualitative inference mechanism can apply these qualitative values to localise fault candidates.

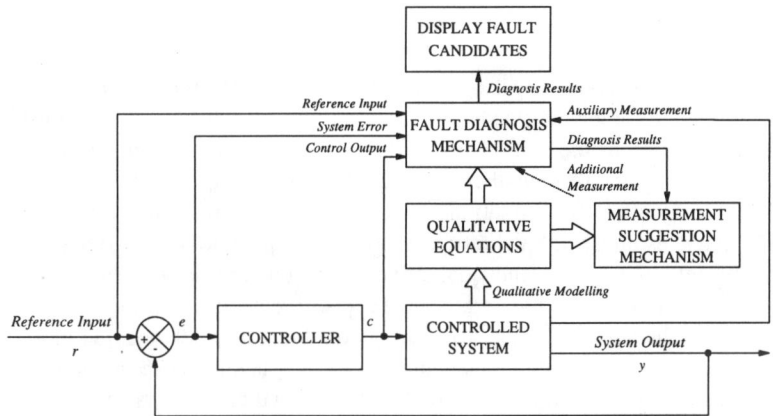

Fig. 6.1 Configuration of the Qualitative Fault Diagnosis Method

The strategy for the fault localisation searches fault locations backwards through various fault hypotheses of particular components to see how a problem might have arisen. When abnormal behaviour is detected, the inference mechanism will firstly assume a component faulty and give the component parameter a qualitative value to represent the assumed fault type. Then, the assumed value and the observed behaviour are inserted into qualitative equations to infer all the qualitative values of unknown variables via operations on qualitative equations. If the qualitative values obtained satisfy all the qualitative equations, then the fault hypothesis is reasonable and thus the component can be seen as a fault candidate. Otherwise the component is not a fault candidate. This process will be reiterated until all the components in a system have been investigated, and then the diagnosis mechanism will report the fault candidates by their locations and fault types.

If the diagnosis result is regarded as requiring refinement, the measurement suggestion mechanism will suggest additional measurements according to the system structure and behaviour. The additional measurements should be measured by users and then inputted manually to the fault diagnosis mechanism as the observed system behaviour. Then, the mechanism will repeat the fault localisation process to generate fault candidates. Theoretically, the more measurements given, the more accurate the diagnosis result obtained.

6.3   QUALITATIVE DESCRIPTORS AND OPERATIONS

In contrast to feedback control tasks, fault diagnosis is performed based on the cause-effect inference rather than accurate computations. As discussed previously, the relationship between the cause and effect is analysed via operations on qualitative equations. Therefore, mathematical qualitative descriptors and operations applicable for the qualitative equations must be defined for the cause-effect inference. In most established qualitative approaches, qualitative descriptors are drawn from the set {+, 0, -, ?}, and qualitative operations are defined in terms of these qualitative values. Here, [+] indicates the positive value in a measurement space, [-] indicates the negative value, [0] denotes the boundary between [+] and [-] values, and [?] expresses the ambiguous value. However, such a coarse division, {+, 0, -}, is not fine enough to distinguish normal and abnormal behaviours for the need of fault diagnosis, since both the normal and abnormal behaviours could be represented by a same qualitative value. Therefore, a finer division for a measurement space and its corresponding qualitative operations are needed.

In order to simply but effectively represent system behaviour and component states for qualitative fault diagnosis, two more qualitative values [1] and [-1] are associated with the set {+, 0 ,-, ?}, where [1] denotes a very big positive value, while [-1] denotes a very big negative value for a variable. For the power variables E and F, [1], [-1], and [0] are used to represent the different abnormal behaviours caused by various system faults, while [+] and [-] are used to represent the normal behaviours. For the component parameters R, C, and I, [1] is used to describe the fault state which will obstruct power delivery, i.e. component blocked or braked, while, [0] is used to describe the fault state which will cause a large amount of power loss, i.e. component leakage or short circuit. Also, [+] and [-] are adopted to describe the normal states for component parameters.

Thus, a set of qualitative operations can be defined according to the implications of these qualitative descriptors. The qualitative operators used here, {+, -, ×, /, =}, are composed from the standard operators for real numbers, and their definitions are shown in Table 6.1.

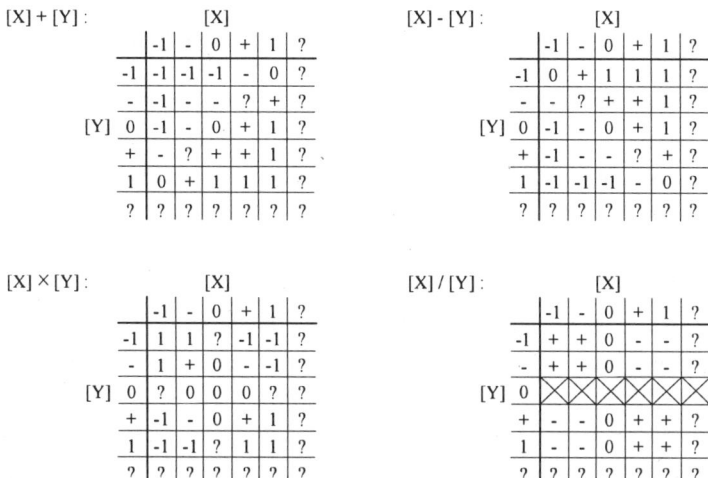

[X] + [Y]:

[Y] \ [X]	-1	-	0	+	1	?
-1	-1	-1	-1	-	0	?
-	-1	-	-	?	+	?
0	-1	-	0	+	1	?
+	-	?	+	+	1	?
1	0	+	1	1	1	?
?	?	?	?	?	?	?

[X] - [Y]:

[Y] \ [X]	-1	-	0	+	1	?
-1	0	+	1	1	1	?
-	-	?	+	+	1	?
0	-1	-	0	+	1	?
+	-1	-	-	?	+	?
1	-1	-1	-1	-	0	?
?	?	?	?	?	?	?

[X] × [Y]:

[Y] \ [X]	-1	-	0	+	1	?
-1	1	1	?	-1	-1	?
-	1	+	0	-	-1	?
0	?	0	0	0	?	?
+	-1	-	0	+	1	?
1	-1	-1	?	1	1	?
?	?	?	?	?	?	?

[X] / [Y]:

[Y] \ [X]	-1	-	0	+	1	?
-1	+	+	0	-	-	?
-	+	+	0	-	-	?
0						
+	-	-	0	+	+	?
1	-	-	0	+	+	?
?	?	?	?	?	?	?

Table 6.1 Qualitative Operations

In Table 6.1 we can see that some uncertain operations, such as "[1] × [0] = [?]", are in contrast to the usual operations for real numbers. These uncertain operations are employed to represent the relationships between the power variables of faulty components, which are not fixed but depend on system structure. Utilising these kinds of uncertain operations guarantees that all the possible relationships between the variables of a faulty component can be considered. This is discussed empirically in the following example.

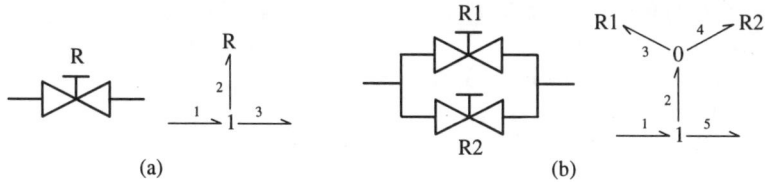

(a)   (b)

Fig. 6.2 Valves and their Bond Graph Models

Fig. 6.2 shows two structures of valves. The qualitative equations of the structure shown in Fig. 6.2 (a) are written as:

$$E1(nT) = E2(nT) + E3(nT),$$
$$F1(nT) = F2(nT) = F3(nT),$$
$$E2(nT) = R \times F2(nT).$$

When the valve R is blocked (R = [1]), the flow passing through the valve and the pressure behind the valve will be zero, but the inlet pressure of R ($E2(nT)$)will be the maximum. Thus, the operations on these qualitative equations will be:

$$E1(nT) = E2(nT) + E3(nT) \qquad \rightarrow \qquad [1] = [1] + [0],$$
$$F1(nT) = F2(nT) = F3(nT) \qquad \rightarrow \qquad [0] = [0] = [0],$$
$$E2(nT) = R \times F2(nT) \qquad \rightarrow \qquad [1] = [1] \times [0].$$

It can be seen from the operation result that the blockage of R will lead the input pressure $E1(nT)$ to the maximum. On the other hand, the qualitative equations of the structure shown in Fig. 6.2 (b) are:

$$E1(nT) = E2(nT) + E5(nT),$$
$$F1(nT) = F2(nT) = F5(nT),$$
$$E2(nT) = E3(nT) = E4(nT),$$
$$F3(nT) = F2(nT) + F4(nT),$$
$$E3(nT) = R1 \times F3(nT),$$
$$E4(nT) = R2 \times F4(nT).$$

Suppose that the valve R1 is blocked (R1 = [1]), but the valve R2 and the input pressure are normal (R2 = [+] and $E1(nT)$ = [+]), then flow passing through R1 ($F3(nT)$) will be zero, and the operations on these equations are:

$$E1(nT) = E2(nT) + E5(nT) \qquad \rightarrow \qquad [+] = [+] + [+],$$
$$F1(nT) = F2(nT) = F5(nT) \qquad \rightarrow \qquad [+] = [+] = [+],$$
$$E2(nT) = E3(nT) = E4(nT) \qquad \rightarrow \qquad [+] = [+] = [+],$$
$$F2(nT) = F3(nT) + F4(nT) \qquad \rightarrow \qquad [+] = [0] + [+],$$
$$E3(nT) = R1 \times F3(nT) \qquad \rightarrow \qquad [+] = [1] \times [0],$$
$$E4(nT) = R2 \times F4(nT) \qquad \rightarrow \qquad [+] = [+] \times [+].$$

Here, the inlet pressures of R1 ($E3(nT)$) and R2 ($E4(nT)$) are equal and remain a normal value ([+]) rather than the maximum value ([1]). It can be seen that the multiplication in the former case is "[1] = [1] × [0]", however, the multiplication in the latter case becomes "[+] = [1] × [0]". This is to say that the relationship between the power variables of a faulty component is not fixed but depends on the system structure, and only the operation "[?] = [1] × [0]" allows all possible relationships can be investigated for fault diagnosis.

6.4   FAULT DETECTION

As in the auto-tuning scheme, the quantitative performance monitoring is again employed to detect the abnormal behaviour caused by system faults. Also, the availability of the fault detection method will not be flawed by system complexity. Fig. 6.3 shows the flowchart of the fault detection process.

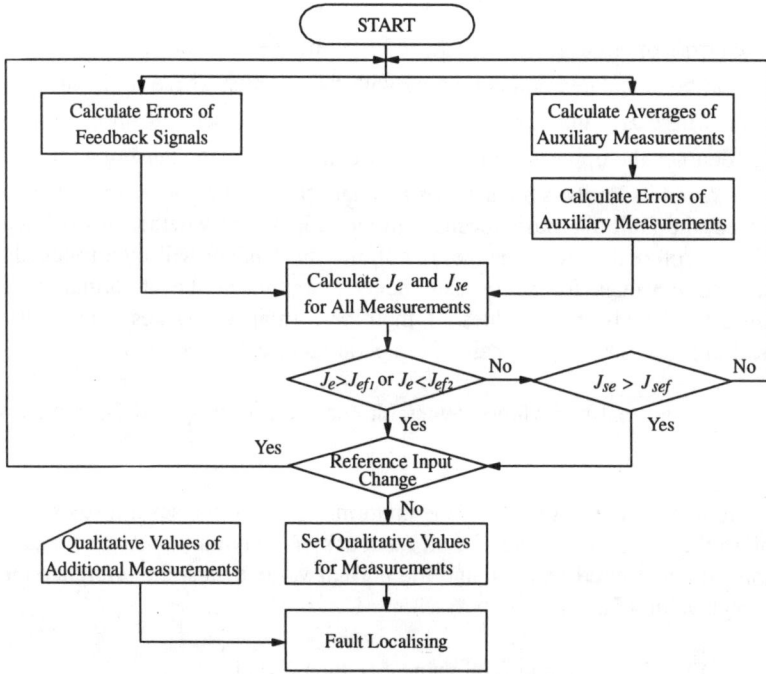

Fig. 6.3 Flowchart of Fault Detection

Currently, the fault diagnosis method is applied to locate the faults occurring in the steady-state, so only the performance indexes J_e and J_{se} (integral error and integral square error) are employed to monitor the steady-state behaviour. The error of a feedback variable can be obtained easily by comparing the observed value to its reference input. However, it is relatively difficult to evaluate the reference value for an auxiliary measurement, which plays a role like the reference input for a feedback measurement, to obtain its error. This is because the reference value is related to the setting of reference inputs, system structure, and component parameters, but a qualitative model does not contain the information about component parameters. A solution for this is to take the average value of past observations of an auxiliary measurement as the reference value. This average value is obtained on-line using the moving exponential window which has been used to get the average value of controller output in Chapter 4. Thus, the error of an auxiliary measurement can be obtained through the equation:

$$\hat{e}(nT) = average\ value(nT) - observed\ value(nT). \tag{6-1}$$

After the errors of all measurements have been obtained, the performance indexes J_e and J_{se} can be calculated and compared with their limits. These limits are given by designers, where J_{ef1} and J_{ef2} denote respectively the upper and lower limits of J_e, and J_{sef} denotes the upper limit of J_{se}. Any component state causing a performance index to exceed its limits is regarded as a system fault. When an abnormal behaviour is detected, the fault diagnosis mechanism will check first whether this behaviour is caused by a reference input change. If not, the mechanism will then proceed to set the qualitative values for all measurements to represent the abnormal behaviour according to the observed values of their performance indexes. Then, the fault localisation process based on qualitative operations can be operated.

The typical abnormal behaviours detectable via the use of J_e and J_{se} are shown in Fig. 6.4.

According to the different types of the abnormal behaviours, several rules are made to explore the qualitative values for all feedback and auxiliary measurements, where X denotes the measured variable and the normal value means the reference input or the average value of a variable:

(1) If $J_e(X) < J_{ef2}(X)$ and normal value ≥ 0, then $X = [1]$.

(2) If $J_e(X) < J_{ef2}(X)$ and normal value < 0, then $X = [0]$.

(3) If $J_e(X) > J_{ef1}(X)$ and normal value ≥ 0, then $X = [0]$.

(4) If $J_e(X) > J_{ef1}(X)$ and normal value < 0, then $X = [-1]$.

(5) If $J_{se}(X) > J_{sef}(X)$ and $J_e(X) < 0$ and normal value ≥ 0, then $X = [1]$.

(6) If $J_{se}(X) > J_{sef}(X)$ and $J_e(X) < 0$ and normal value < 0, then $X = [0]$.

(7) If $J_{se}(X) > J_{sef}(X)$ and $J_e(X) > 0$ and normal value ≥ 0, then $X = [-1]$.

(8) If $J_{se}(X) > J_{sef}(X)$ and $J_e(X) > 0$ and normal value < 0, then $X = [0]$.

(9)   If $J_{se}(X) > J_{sef}(X)$ and $J_e(X) = 0$, then do not set any values.

(10)   Else $X = [+]$ when normal value ≥ 0, or $X = [-]$ when normal value < 0.

Since the error used in this approach is defined as *error = set-point - observed value*, a negative error denotes that the measured value of a variable is bigger than its normal value, while a positive error denotes that the measured value is smaller then the normal one. Thus, rules (1) and (2) can be used to set the qualitative value for the behaviours of (a-1) and (b) in Fig. 6.4, where the measured values are much bigger than their normal values; whereas rules (3) and (4) can be used to set the qualitative value for the behaviours of (a-2) and (c), where the measured values are smaller than their normal values. Besides, rules (5) and (6) set the qualitative values

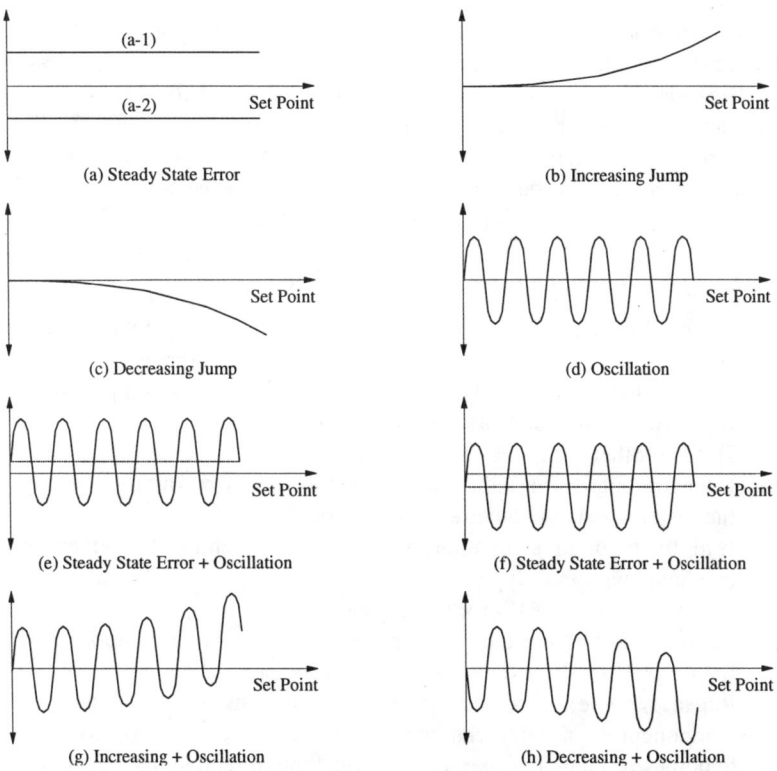

Fig. 6.4 Typical Abnormal Behaviours of a System

for the behaviours of (e) and (g) in Fig. 6.4, where the error sum is negative so that the measured values are bigger than their normal values. Similarly, rules (7) and (8) are used to set the values for the behaviour of (f) and (h). However, it is very difficult to use a qualitative value to describe a pure oscillation as shown in Fig. 6.4 (d). Therefore, the fault diagnosis mechanism will not set any values for the variable which represents a pure oscillation. In this case, the variable with a pure oscillation will be seen as having an unknown value, and fault candidates will be localised according to other measurements. After all, if J_e and J_{se} of a variable are both within the acceptable ranges, the qualitative value of the variable will be set as [+] or [-], which denotes a normal value.

6.5   FAULT LOCALISATION

After an abnormal behaviour is detected and the qualitative values for all measurements are set, the fault diagnosis mechanism will turn to localise system faults. The possible fault candidates of system components, controllers, power drivers, and sensors are localised via the following automatic sequence. The basic idea underlying the fault localisation sequence is that a fault hypothesis is recognised when it can be linked to observations reasonably via qualitative equations.

1) *Localising Fault Candidates of System Components:*

   a)   Insert all the qualitative values of measurements into qualitative equations.
   b)   Assume that a component is faulty and set its fault type as [1].
   c)   Set the flow of the faulty component as [0], since a component with the fault type [1] will have its flow tending to zero.
   d)   Hold all other components as normal ([+],or [-]).
   e)   Infer the qualitative values for all unknown power variables according to the inserted data via qualitative operations.
   f)   If all the qualitative equations are recognised, then the component is a fault candidate with fault type [1].
   g)   Set the fault type as [0] for this component.
   h)   Set the effort of the faulty component as [0], since a component with the fault type [0] will have its effort tending to zero.
   i)   Repeat steps e). If all the qualitative equations are recognised, then the component is a fault candidate with fault type [0]. If both the fault hypotheses cannot make all the qualitative equations recognised, the component is not a fault candidate.
   j)   Reiterate steps b) to i) until all components have been investigated.

2) *Localising Fault Candidates of Controllers:*

    a)   Insert all the qualitative values of measurements into qualitative equations.

    b)   Hold all system components as normal.

    c)   Map a controller output into its qualitative value, and then insert the value into the controlled power variable.

    d)   Infer the qualitative values for all unknown power variables according to the inserted data via qualitative operations.

    e)   If all the qualitative equations are recognised, then the controller is a fault candidate. Otherwise, it's not a fault candidate.

    f)   Reiterate steps c) to e) until all controllers have been investigated.

3) *Localising Fault Candidates of Power Drivers:*

    a)   Insert all the qualitative values of measurements into qualitative equations.

    b)   Hold all components as normal.

    c)   If a controller is not regarded a fault candidate via Step 2), than assume that the power driver governed by the controller has a deviated value from the controller output. For example, if a controller output is [1] or [-1], then assume that power driver outputs the value [0].

    d)   Insert the fault hypothesis into the controlled power variable.

    e)   Infer the qualitative values for all unknown power variables according to the inserted data via qualitative operations.

    f)   If all the qualitative equations are recognised, then the controller is a fault candidate. Otherwise, it's not a fault candidate.

    g)   Reiterate steps c) to f) until all power drivers have been investigated.

4) *If no fault candidate can be found through the above three steps, then go to Step 5) to localise sensors' faults. Otherwise, the fault localisation sequence is finished.*

5) *Localising Fault Candidates of Sensors:*

    a)   Assume that the value of a variable which a sensor measures is in contrast to the observed one. For example, If a sensor indicates a measurement [1] on a variable, then set the variable as [0].

    b)   Insert the fault hypothesis into the qualitative equations, and then repeat Steps 1) to 3) to find fault candidates.

    c)   If some fault candidates can be found by this assumption, then the sensor is a fault candidate, and the fault candidates found through Steps 1) to 3) are also recognised. Otherwise, the sensor is not a fault candidate.

    d)   Reiterate steps a) to c) until all sensors have been investigated.

The method of inferring qualitative values for unknown variables is practised by the following steps:

1) Set the past states of C and I components as [+], and then insert the observed values and the fault hypothesis into the qualitative equations. However, if a setting causes an ambiguity, such as $[+] \times ([+] - [+]) = [?]$, then set the past state as a smaller value to avoid the ambiguity. For example, in this case, the past state can be set as [0], and then the operation will be "$[+] \times ([+] - [0]) = [+]$".

2) Find the equations which have only one unknown variable, or the equations which contain only "=" operator and have at least one variable whose value is already known. Then, evaluate the unknown variables through qualitative operations. If the value of an unknown variable cannot be obtained certainly (this can be caused by the uncertain operations, such as $[1] \times [0] = [?]$), then the variable remains as an unknown one, until its value can be obtained from other equations.

3) Insert the newly obtained values into qualitative equations, and repeat Step 2) until all qualitative equations have been determined.

4) If an equation is found violated and one of its variables is related to a C or I component, then the past state of the C or I component will be changed to other values by the sequence "$[+] \rightarrow [0] \rightarrow [1] \rightarrow [-] \rightarrow [-1]$". For example, if the past state is [+], then it will be changed to [0]; if the past state is [0], then it will be changed to [1]. Then, repeat Steps 2) and 3) to re-evaluate the unknown variables to see whether all the equations can be recognised. There are two reasons for this. Firstly, the value [+] cover a quite wide interval. It could indicate either a value near [0] or a value near [1]. The accurate information lost here could cause some possible situations to be eliminated from the inference process. Secondly, all the past states of C and I components have been set as [+] in Step 1). However, some past states could be actually [-] or [-1]. These situations should also be investigated. The expediency of changing the past states for C and I components is employed to guarantee that all possible system situations can be considered.

6.5.1   An Example

The SISO coupled tanks liquid level control rig used in Chapter 4 is employed to help explain the fault localisation process. The schematic diagram of the system is shown again in Fig. 6.5, and its qualitative equations are established in Eqs. (6-2) to (6-26).

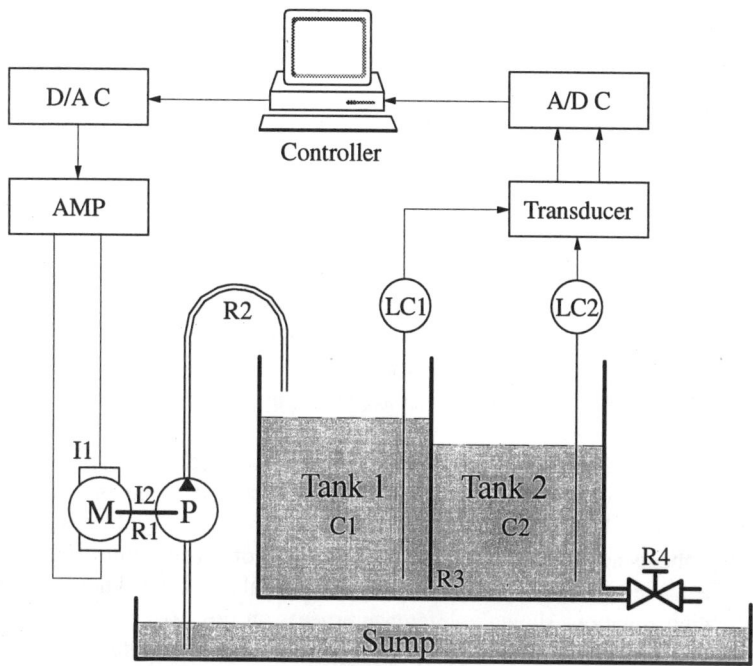

Fig. 6.5 Schematic Diagram of the SISO Coupled Tanks Liquid Level Control Rig

$$E1(nT) = E2(nT) + E3(nT), \tag{6-2}$$
$$F1(nT) = F2(nT) = F3(nT), \tag{6-3}$$
$$E2(nT) = I1 \times (F2(nT) - F2((n-1)T), \tag{6-4}$$
$$E3(nT) = F4(nT), \tag{6-5}$$
$$F3(nT) = E4(nT), \tag{6-6}$$
$$E4(nT) = E5(nT) + E6(nT) + E7(nT), \tag{6-7}$$
$$F4(nT) = F5(nT) = F6(nT) = F7(nT), \tag{6-8}$$
$$E5(nT) = I2 \times (F5(nT) - F5((n-1)T), \tag{6-9}$$
$$E6(nT) = R1 \times F6(nT), \tag{6-10}$$
$$E7(nT) = E8(nT), \tag{6-11}$$
$$F7(nT) = F8(nT), \tag{6-12}$$
$$E8(nT) = E9(nT) + E10(nT), \tag{6-13}$$
$$F8(nT) = F9(nT) = F10(nT), \tag{6-14}$$
$$E9(nT) = R2 \times F9(nT), \tag{6-15}$$
$$E10(nT) = 0, \tag{6-16}$$
$$E11(nT) = E12(nT), \tag{6-17}$$
$$F10(nT) = F11(nT) + F12(nT), \tag{6-18}$$
$$F11(nT) = C1 \times (E11(nT) - E11((n-1)T), \tag{6-19}$$

$$E12(nT) = E13(nT) + E14(nT), \tag{6-20}$$
$$F12(nT) = F13(nT) = F14(nT), \tag{6-21}$$
$$E13(nT) = R3 \times F13(nT), \tag{6-22}$$
$$E14(nT) = E15(nT) = E16(nT), \tag{6-23}$$
$$F14(nT) = F15(nT) + F16(nT), \tag{6-24}$$
$$F15(nT) = C2 \times (E15(nT) - E15((n-1)T), \tag{6-25}$$
$$E16(nT) = R4 \times F16(nT). \tag{6-26}$$

Suppose that there are two sensors in this system: one is the auxiliary sensor used to measure Tank 1 liquid level, the other is the feedback sensor used to measure Tank 2 liquid level. Here, we cite an example where both the sensors indicate the liquid levels at increasing jumps. Thus, the variables $E11(nT)$ (Tank 1 liquid level) and $E15(nT)$ (Tank 2 liquid level) will be set as [1] first by the inference mechanism.

After an abnormal behaviour is detected, the inference mechanism will investigate the system components firstly. Thus, R4 is now assumed to be blocked and its parameter is given by [1], the flow of R4 ($F16$) is given by [0], the parameters of other components are all set as [+], and the past states of C and I components are set as [+]. Then, the inference mechanism will go to evaluate the unknown variables. The following equations show the inference process on Eqs. (6-17) to (6-26), where the serial number shown in front of the equations represent the inference sequence. First, the unknown variables in the equations numbered 1) can be obtained, since these equations have either only one unknown variable or only "=" operator. With these newly obtained values, the variables in the equations numbered 2) can then be evaluated. Similarly, all other variables can be evaluated sequentially.

1) $E11(nT) = E12(nT)$ $\qquad \rightarrow \quad [1] = [1],$
4) $F10(nT) = F11(nT) + F12(nT)$ $\qquad \rightarrow \quad [+] = [+] + [+],$
1) $F11(nT) = C1 \times (E11(nT) - E11((n-1)T) \quad \rightarrow \quad [+] = [+] \times ([1] - [+]),$
5) $E12(nT) = E13(nT) + E14(nT)$ $\qquad \rightarrow \quad [1] = [+] + [1],$
3) $F12(nT) = F13(nT) = F14(nT)$ $\qquad \rightarrow \quad [+] = [+] = [+],$
4) $E13(nT) = R3 \times F13(nT)$ $\qquad \rightarrow \quad [+] = [+] \times [+],$
1) $E14(nT) = E15(nT) = E16(nT)$ $\qquad \rightarrow \quad [1] = [1] = [1],$
2) $F14(nT) = F15(nT) + F16(nT)$ $\qquad \rightarrow \quad [+] = [+] + [0],$
1) $F15(nT) = C2 \times (E15(nT) - E15((n-1)T)$ $\quad \rightarrow \quad [+] = [+] \times ([1] - [+]),$
2) $E16(nT) = R4 \times F16(nT)$ $\qquad \rightarrow \quad [1] = [1] \times [0].$

It can be seen that the above equations are satisfied. Moreover, since the known qualitative values in the equations Eqs. (6-2) to (6-16) are all [+], it can be inferred that all their unknown variables will also be [+]. Thus, these equations can be recognised. Therefore, the assumption, R4 blocked, is recognised. In other words, the increasing jump of the liquid level could be caused by the blocked outlet tap.

Next, R4 is now assumed to be leaking. Thus, the inference on Eqs. (6-23) to (6-26) will be:

1) $E14(nT) = E15(nT) = E16(nT)$ $\rightarrow$ $[1] = [1] = [0]$,
2) $F14(nT) = F15(nT) + F16(nT)$ $\rightarrow$ $[?] = [+] + [?]$,
1) $F15(nT) = C2 \times (E15(nT) - E15((n-1)T)$ $\rightarrow$ $[+] = [+] \times ([1] - [+])$,
2) $E16(nT) = R4 \times F16(nT)$ $\rightarrow$ $[0] = [0] \times [?]$.

Obviously, the equation $[1] = [1] = [0]$ is violated, and no situation can make this equation reasonable. Therefore, R4 leakage cannot be proved as a fault candidate. That is, it is impossible for the leaking outlet tap to cause an increasing jump of the liquid levels. Through this cause-effect inference, all the components can be investigated and the fault candidates of system components can be obtained as R4 and C2 (Tank 2) blocked.

Then, the inference mechanism will turn to consider the controller faults. It is assumed that the controller in this feedback control system is normal, so the controller output must be reduced to cope with the increasing jumps of the liquid levels. Thus, the controller output can be mapped to $[0]$, and the value of the controlled variable $E1(nT)$ is then given by $[0]$. Meanwhile, all the components in this system are set as normal. Thus, all the unknown variables can be inferred as follows:

10) $E1(nT) = E2(nT) + E3(nT)$ $\rightarrow$ $[0] = [-1] + [1]$,
12) $F1(nT) = F2(nT) = F3(nT)$ $\rightarrow$ $[-1] = [-1] = [1]$,
11) $E2(nT) = I1 \times (F2(nT) - F2((n-1)T)$ $\rightarrow$ $[-1] = [+] \times ([-1] - [0])$,
9) $E3(nT) = F4(nT)$ $\rightarrow$ $[1] = [1]$,
11) $F3(nT) = E4(nT)$ $\rightarrow$ $[1] = [1]$,
10) $E4(nT) = E5(nT) + E6(nT) + E7(nT)$ $\rightarrow$ $[1] = [+] + [1] + [1]$,
8) $F4(nT) = F5(nT) = F6(nT) = F7(nT)$ $\rightarrow$ $[1] = [1] = [1] = [1]$,
9) $E5(nT) = I2 \times (F5(nT) - F5((n-1)T)$ $\rightarrow$ $[+] = [+] \times ([1] - [+])$,
9) $E6(nT) = R1 \times F6(nT)$ $\rightarrow$ $[1] = [+] \times [1]$,
9) $E7(nT) = E8(nT)$ $\rightarrow$ $[1] = [1]$,
7) $F7(nT) = F8(nT)$ $\rightarrow$ $[1] = [1]$,
8) $E8(nT) = E9(nT) + E10(nT)$ $\rightarrow$ $[1] = [1] + [0]$,
6) $F8(nT) = F9(nT) = F10(nT)$ $\rightarrow$ $[1] = [1] = [1]$,
7) $E9(nT) = R2 \times F9(nT)$ $\rightarrow$ $[1] = [+] \times [1]$,

1) $E10(nT) = 0$ $\rightarrow$ $[0] = [0]$,
1) $E11(nT) = E12(nT)$ $\rightarrow$ $[1] = [1]$,
5) $F10(nT) = F11(nT) + F12(nT)$ $\rightarrow$ $[1] = [+] + [1]$,
1) $F11(nT) = C1 \times (E11(nT) - E11((n-1)T)$ $\rightarrow$ $[+] = [+] \times ([1] - [+])$,
6) $E12(nT) = E13(nT) + E14(nT)$ $\rightarrow$ $[1] = [1] + [1]$,
4) $F12(nT) = F13(nT) = F14(nT)$ $\rightarrow$ $[1] = [1] = [1]$,
5) $E13(nT) = R3 \times F13(nT)$ $\rightarrow$ $[1] = [+] \times [1]$,
1) $E14(nT) = E15(nT) = E16(nT)$ $\rightarrow$ $[1] = [1] = [1]$,
3) $F14(nT) = F15(nT) + F16(nT)$ $\rightarrow$ $[1] = [+] + [1]$,
1) $F15(nT) = C2 \times (E15(nT) - E15((n-1)T)$ $\rightarrow$ $[+] = [+] \times ([1] - [+])$,
2) $E16(nT) = R4 \times F16(nT)$ $\rightarrow$ $[1] = [+] \times [1]$.

It can be seen that the second equation $[-1] = [-1] = [1]$ is violated. That is, the controller which has reduced its output to cope with the increasing liquid levels is not a fault candidate. In contrast, if the controller output is increased in this situation, then the controller can be seen as a fault candidate.

Then, the inference mechanism will turn to investigate the power driver. Since the controller has been regarded as normal, the power driver regulated by the controller is assumed faulty and its output is thus given by $[1]$ which is opposite to the controller output $[0]$. Thus, the controlled variable $E1(nT)$ is set as $[1]$. It can be seen from the above inference that if $E1(nT) = [1]$, then the unknown variables in the first three equations will be obtained as:

$E1(nT) = E2(nT) + E3(nT)$ $\rightarrow$ $[1] = [+] + [1]$,
$F1(nT) = F2(nT) = F3(nT)$ $\rightarrow$ $[1] = [1] = [1]$,
$E2(nT) = I1 \times (F2(nT) - F2((n-1)T)$ $\rightarrow$ $[+] = [+] \times ([1] - [+])$,

Thus, all the qualitative equations will be recognised. In other words, the power driver is a fault candidate. This means that the increasing jumps of the liquid levels may be caused by a faulty power driver which does not follow the controller output.

Finally, let us consider the possible sensor faults. Since several fault candidates have been found through the above inference according to the measurements, these measurements are regarded as reasonable. Therefore, the sensors which supply these reasonable measurements are regarded as normal and the fault localisation process will be finished here. However, if the sensors provide an unreasonable measurement, such as Tank 1 liquid level decreasing fast and Tank 2 liquid level increasing fast ($E11(nT) = [0]$ and $E15(nT) = [1]$), then no fault candidates can be found through the above inference. In this case, the inference mechanism will assume that Tank 1 sensor is faulty and set $E11(nT)$ as [1] (which is opposite to the actual measurement [0]), and then go back to localise fault candidates. We can see that the measurements now are the same as in the above case ($E11(nT) = [1]$ and $E15(nT) = [1]$). Therefore, several fault candidates can be found from the above discussion. This is to say that Tank 1 sensor is a fault candidate. Next, Tank 2 sensor is assumed to be faulty and the measurement of Tank 2 is given by [0]. Thus, the measurements will be $E11(nT)$ = [0] and $E15(nT) = [0]$. In this circumstance, several fault candidates can also be found. Therefore, Tank 2 sensor is also a fault candidate.

In this SISO coupled tanks example, no ambiguity has been caused by qualitative operations to obstruct the cause-effect inference for fault diagnosis. One reason for this is that the use of adequate qualitative descriptors and operations avoids most opportunities to cause ambiguity. The other reason is that all the variables in this example have only one possible normal value [+] rather than [+] or [-], so ambiguous equations, such as [?] = [+] + [-], did not appear in the inference process. However, if some ambiguous values do exist in the inference process, then the inference mechanism will ignore the equations containing the ambiguous values and investigate other equations. If all the explicit equations which contain no ambiguous values are satisfied, then the fault hypothesis can be seen acceptable. In contrast, if some explicit equations are violated, then the fault hypothesis is not recognised.

6.5.2   Efficiency Improvement

Usually, an engineering system is constructed with many components. Investigating all the components one by one to locate fault candidates will take too much time for efficient on-line fault diagnosis. Accordingly, an efficiency improvement method is developed to reduce the number of process investigations. The basic idea for this method is to decompose a system into several segments, where each component with an identical fault stage within a segment will cause the system to have similar abnormal behaviour. In other words, once a component has been identified as a fault candidate, other components in the same segment are fault candidates as well. Thus, fault candidates can be localised without checking every component individually so that the diagnosis efficiency can be improved.

Through analysing a system structurally, a number of significant "landmarks" can be found. With these landmarks, the physical system can be decomposed into several segments, where each component in a segment can be regarded as having the same effect on a system. Qualitative equations used to represent bond graph models contain exactly enough structural information to support finding landmarks for physical systems. An example shown in Fig. 6.6 is employed to help explain how to find the landmarks with which to decompose a bond graph model. Here, the typical physical interconnections of an engineering system are analysed, and then the significant landmarks are found through analysing its qualitative equations.

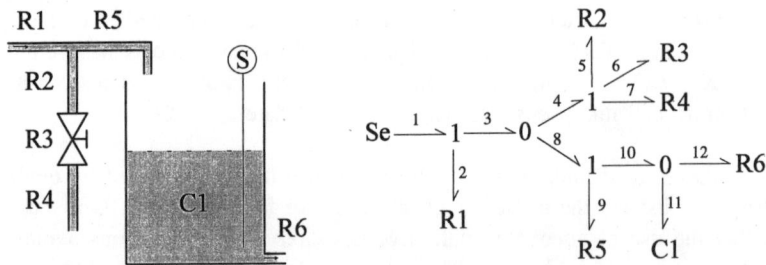

Fig. 6.6 A Liquid Pipe and Tank Rig and its Bond Graph Model

1) *Serial Junction:*

A serial junction is a common flow junction, which provides only one path for power delivery. Any component fault in the path will obstruct power delivery. For example, in this case, components R2, R3, and R4 are connected with a serial junction. Thus, any blockage of R2, R3, or R4 will stop the flow going through these pipes and valve, while any leakage of these components will decrease the flow passing through R5. However, there are still some differences. If the leakage occurs in R2, then the flow passing through R3 will decrease. But, if the leakage occurs in R4, then the flow passing through R3 will increase. In this situation, a sensor located on R3 is necessary to distinguish this difference in a qualitative diagnosis. Thus, it can be concluded that a similar fault happening to any components connected with a serial junction can be assumed to have the same effect on a system when there are no sensors used to observe the actual system behaviour. In other words, components connected with each others by a serial junction without sensors between them can be regarded as one segment. Therefore, serial junction is not a landmark which can be used to decompose bond graph models.

2) *Parallel Junction:*

A parallel junction is a common effort junction, which provides more than one path for power delivery. Components connected with a parallel junction are located on different power paths, so that a similar fault happening to the components could cause different effects on a system. For example, R2 and R5 are connected with a parallel junction. If R2 is blocked, the flow passing through R2 will decrease and the flow passing through R5 will increase. On the other hand, if R5 is blocked, then the flow going through R2 will increase and the flow passing through R5 will decrease. Thus, components connected with a parallel junction cannot be seen as a segment, so a parallel junction can be used as a landmark.

3)   *Complete Interaction:*

Components with a complete interaction interact with each other in both aspects of effort and flow. System power can be delivered forwards or backwards through a complete interaction. In the example, R1, R2, R3, R4 and R5 interact with each other by a complete interaction. According to the reason discussed previously, the components interact with each other with a complete interaction and without sensors and parallel junctions between them can be assumed to have the same effect on a system, and can be seen as a segment. Thus, the complete interaction cannot be used as a landmark.

4)   *Incomplete Interaction:*

In such an interaction, a component interacts with others in either effort or flow. System power can only be delivered in a single direction. In this example, R5 has an incomplete interaction with C1 and R6, where the liquid level of C1 cannot influence the effort and flow of R5. System power can be delivered from R5 to C1 and R6, but cannot be delivered from C1 to R5. Here, the bond 10 is called an incomplete interaction, and variation behind the bond 10 will not affect the components located in front of the bond. That is, components connected with an incomplete interaction cannot be regarded as a segment, and the incomplete interaction can be seen as a landmark.

5)   *Non-interaction:*

Two non-interacted components have no interaction with each other, so their behaviours will not affect each other. Thus, components connected with a non-interaction are not in a segment, and a non-interaction can be seen as a landmark.

6)   *Starting Point:*

System power is fed into a system from starting points, so a starting point must be the beginning of a segment. Thus, starting point can be seen as a landmark.

7) *Ending Point:*

Ending point is the end of the power delivery in a system. In this example, the power paths end in R4 and R6. Thus, such a point must be an end of a segment, and can be regarded as a landmark.

8) *Measurement Point:*

Measurement points are where sensors are located. As discussed in the point 1), the use of a sensor can identify a fault occurring before or behind a measurement point. Therefore, behaviours of the components located before or behind such a point can be analysed separately. Thus, a measurement point can be seen as a landmark.

In conclusion, parallel junction, incomplete interaction, non-interaction, starting point, ending point, and measurement point are employed as the landmarks to decompose a bond graph model.

How to identify the landmarks from a set of qualitative equations is now explained using the same pipes and tank system. Here, the qualitative equations of this system are written as follows:

$$E1(nT) = E2(nT) + E3(nT), \tag{6-27}$$
$$F1(nT) = F2(nT) = F3(nT), \tag{6-28}$$
$$E2(nT) = R1 \times F2(nT), \tag{6-29}$$
$$E3(nT) = E4(nT) = E8(nT), \tag{6-30}$$
$$F3(nT) = F4(nT) + F8(nT), \tag{6-31}$$
$$E4(nT) = E5(nT) + E6(nT) + E7(nT), \tag{6-32}$$
$$F4(nT) = F5(nT) = F6(nT) = F7(nT), \tag{6-33}$$
$$E5(nT) = R2 \times F5(nT), \tag{6-34}$$
$$E6(nT) = R3 \times F6(nT), \tag{6-35}$$
$$E7(nT) = R4 \times F7(nT), \tag{6-36}$$
$$E8(nT) = E9(nT) + E10(nT), \tag{6-37}$$
$$F8(nT) = F9(nT) = F10(nT), \tag{6-38}$$
$$E9(nT) = R5 \times F9(nT), \tag{6-39}$$
$$E10(nT) = 0, \tag{6-40}$$
$$E11(nT) = E12(nT), \tag{6-41}$$
$$F10(nT) = F11(nT) + F12(nT), \tag{6-42}$$
$$F11(nT) = C1 \times (E11(nT) - E11((n-1)T)), \tag{6-43}$$
$$E12(nT) = R6 \times F12(nT). \tag{6-44}$$

Firstly, a sensor is given by the user to measure $E11$, so Bond-11 is a measurement point and a landmark. Secondly, it can be found from these equations that $E1$ and $F1$ only appear on the left hand side of "=", which means that power is fed into the system from Bond-1. Therefore, Bond-1 is a starting point of the system and is a landmark. Thirdly, Eqs. (6-30) and (6-31), and (6-41) and (6-42) show respectively two common effort junctions. Thus, the components located on each side of "=" in Eqs. (6-31) and (6-42) can be identified as being in different segments. Next, Eqs. (6-40) to (6-42) show that Bond-10 is an incomplete interaction so that it is a landmark. Finally, it can be found that $F7$ and $F12$ only appear on the right hand side of "=", which means that the components connected on Bond-7 and Bond-12 only receive power but do not transport the power to any other component. Therefore, the components connected to Bond-7 and Bond- 12 are the end points of the system and are landmarks.

Ultimately, the system is decomposed into the segments: {R1}, {R2, R3, R4}, {R5}, {C1}, and {R6}. For a fault diagnosis inference, the number of process investigations processes can be reduced from 7 to 5. Thus, the fault localisation efficiency can be improved. Moreover, a bond graph model can be decomposed at the modelling stage, so it will not impair the efficiency for on-line fault diagnosis.

6.6   MEASUREMENT SUGGESTION

A generic inference methodology based on system structure analysis is developed here to suggest additional measurements automatically to refine the initial fault diagnosis result:

1)   Search forwards through every power path from the starting point of each path to find the first measurement point where a sensor assesses a fault state.

2)   If the fault state is an increasing jump, then the components between the starting point and the measurement point can be judged to be normal. The reason is that only normal components can deliver power through the path so that the energy stored or lost at the measurement point can increase. Thus, fault candidates of system components can only be found behind the measurement point, and it is not necessary to locate additional measurements between the starting point and the measurement point. Then, search forwards through the path from the measurement point to find the first incomplete interaction point. If such a point is found, then the power variable of the component in front of this point is suggested as an additional measurement. The reason is that, usually, the behaviour of this component is the easiest to be observed. For example, in the

example illustrated in Fig. 6.6, the flow of R5 is very easy to observe. If no incomplete interaction can been found, then the power variable of the neighbouring component behind the measurement point is suggested as an additional measurement. However, if the chosen component has already been measured, then the variable of the component next to the already measured one is an additional measurement.

3) If the fault state is a decreasing jump, then the components between the starting point and the measurement point could be faulty because a decreasing jump means that some of the components obstruct the power delivery. Thus, additional measurements can be located between the starting and the measurement points to refine the fault diagnosis result. Thus, search backwards through the path from the measurement point to find the first incomplete interaction. Then , the variable of the component in front of the incomplete interaction will be an additional measurement. If there is no incomplete interaction, then the variable of the neighbouring component in front of the measurement point will be an additional measurement.

4) Search backwards through the path from the measurement point to find the first parallel junction. If there is no parallel junction, then the measurement suggestion process is finished. However, if a parallel junction is found, then search forwards through other paths connected to the parallel junction from this junction to find the first incomplete interaction. Thus, the variable of the component in front of the incomplete interaction will be an additional measurement. If there is no incomplete interaction, then the variable of the neighbouring component behind parallel junction is an additional measurement. If the additional measurement found in this step has already been chosen by the previous two steps, then the variable of the component next to the already chosen one is an additional measurement.

5) If a suggested component is connected to the system with a common flow junction, then the additional measurement should be its flow variable. If a suggested component is connected to the system with a common effort junction, then the additional measurement should be its effort variable. The reason for doing so is to ensure the values of more variables and to avoid possible ambiguity.

After this, the user is required to input the observations of the additional measurement by qualitative values. Then, the fault diagnosis mechanism will repeat the fault localisation process again to localise the fault candidates. Usually, the more the measurements the narrower the fault localising possibility. Thus, the fault diagnosis result obtained from fewer measurements can be refined by using the additional measurements. Here, the qualitative values of the additional measurements are determined by the user. When the normal value of an additional measurement is [+], then [1] can be used to denote a value bigger than the normal one, [0] denotes the value smaller than the normal one, [+] indicates a normal value, and [-] can be used to denote a vector in the opposite direction to the normal value. When the normal value of an additional measurement is [-], then [-1], [0], [-], and [+] can be used respectively to represent measurements which are bigger, smaller, normal, and in opposite direction in comparison with the normal value.

Here, the system shown in Fig, 6.6 is employed again to help explain this measurement suggestion method. Assume that there is a sensor used to measure the liquid level of the tank, and the liquid level is measured decreasing very fast with the controller being normal. Then, the initial fault candidates can be obtained as:

> component leakage: R2, R3, R4, R6,
> component leakage or block: R1, R5, C1,
> power driver fault: Driver-1.

Firstly, the measurement suggestion mechanism will search forwards through the paths Bond-1, 2, 3, 4, 5, 6, 7, and Bond-1, 2, 3, 8, 9, 10, 11, 12, to find the first measurement where an abnormal state is assessed. Thus, the sensor located on C1 is found denoting an abnormal state [0]. Secondly, according to the fault state, the mechanism will search backwards from the measurement point through the path Bond-11, 10, 9, 8, 3, 2, 1, to find the first incomplete interaction. Thus, Bond-10 is found, and then the component in front of Bond-10 (R5) is suggested to be measured. Next, the mechanism will search backwards from the measurement point through the path Bond-11, 10, 9, 8, 3, 2, 1, to find the first parallel junction. Thus, the junction connecting Bond-3, 4, and 8 is found. Then, the mechanism will search forwards from the junction through the path Bond-4, 5, 6, 7, to find the incomplete interaction. However, there is no incomplete interaction to be found. Thus, the neighbouring component behind the parallel junction (R2) is suggested to be measured. Since R5 and R2 are connected to the system with common flow junctions, their flow variables F9 and F5 are suggested as the additional measurement. Then, the user is required to observe the flows of R2 and R5 and input the observations.

If the flow of R2 is increasing and the flow of R5 is normal, then their qualitative values can be given by [1] and [+] respectively. Thus, the fault diagnosis result can be refined by re-localising fault candidates with two more measurements, and the refined result is as follows:

          component leakage: R2, R3, R4, R6,
          component leakage or block: C1.

It can be seen that the initial fault candidates R1, R5 and Driver-1 are rejected. This result is quite reasonable, since the system power can be delivered to R2 and R5 only when R1, R5, and Driver-1 are normal.

6.7 CASE STUDY

The implementation of the fault diagnosis method is illustrated by the same MIMO coupled tanks liquid level control rig used in previous chapters. The schematic diagram of the MIMO system is shown again in Fig. 6.7 (see Fig. 4.27 for its bond graph model, and Eqs. (4-52) to (4-91) for its qualitative equations).

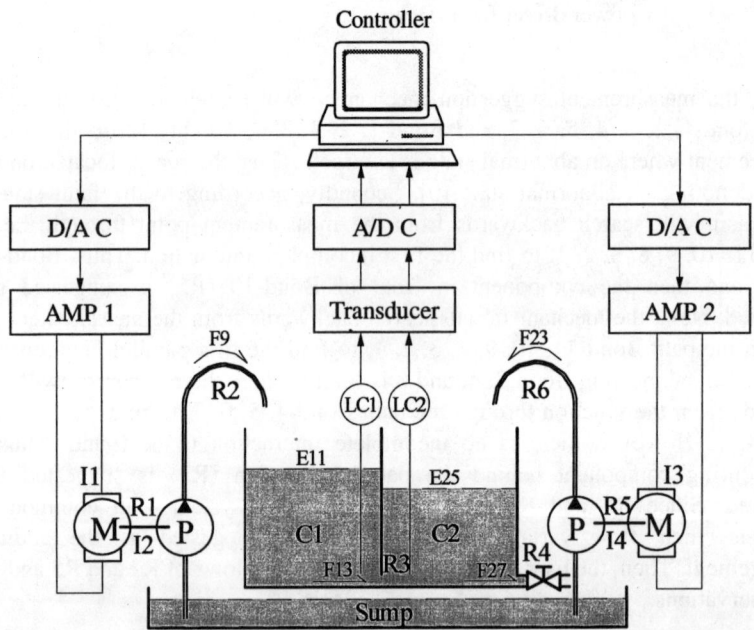

Fig. 6.7 Schematic Diagram of the MIMO Coupled Tanks Liquid Level Control Rig

All the following experiments began with the components being normal. After the system achieved steady-state, some faults were imposed onto the system manually. When an abnormal behaviour was detected, the controllers stopped regulating the system and began to localise system faults. Fault candidates were then represented by the symbols of system components and their fault types. The measurement suggestions were given together with the fault candidates, and users could input the additional measurements manually to refine the diagnosis result. If no more additional measurement could be suggested, the fault diagnosis mechanism would also give users this information. The symbols used to represent fault candidates are:

Pump-1	: the pump pumping water into Tank 1,
Pump-2	: the pump pumping water into Tank 2,
Motor-1	: the motor actuating Pump-1,
Motor-2	: the motor actuating Pump-2,
Driver-1	: the driver driving Motor-1,
Driver-2	: the driver driving Motor-2,
Controller-1	: the controller regulating Driver-1,
Controller-2	: the controller regulating Driver-2,
Sensor-1	: the sensor used to measure Tank 1 liquid level,
Sensor-2	: the sensor used to measure Tank 2 liquid level,
C1	: Tank 1,
C2	: Tank 2,
I1	: armature inductance of Motor-1,
I2	: inertia of Motor-1 and Pump-1,
I3	: armature inductance of Motor-2,
I4	: inertia of Motor-2 and Pump-2,
R1	: axial friction of Motor-1 and Pump-1,
R2	: inlet pipe of Tank 1,
R3	: orifice between Tank 1 and Tank 2,
R4	: discharge tap of Tank 2,
R5	: axial friction of Motor-2 and Pimp-2
R6	: inlet pipe of Tank 2.

The setting of the scaling factors of the hybrid qualitative and quantitative controllers were:

Controller-1: $SF_e = 0.5$, $SF_{ec} = 30$, $SF_o = 0.5$, $\delta = 0.9$,
Controller-2: $SF_e = 0.5$, $SF_{ec} = 30$, $SF_o = 1.0$, $\delta = 0.9$.

When steady-state error exceeded ±7.5% of the measurement space over ten sampling periods, the behaviour was regarded as being caused by system faults.

Thus, the performance criteria were obtained as: $J_{ef1} = 7.5\% \cdot 2048 \cdot 10 \approx 1500$, $J_{ef2} = -7.5\% \cdot 2048 \cdot 10 \approx -1500$, and $J_{sef} = (7.5\% \cdot 2048)^2 \cdot 10 \approx 225000$. Further, the measurement suggestion mechanism was limited to suggest not more than three additional measurements at once. The rules used in these experiments to map the controller outputs into qualitative values were as follows:

1) If $\sum\limits_{k=0}^{9} OC((n-k)T) > 0$, $|OC(nT)| > 10\%$ controller output range, and $A(nT) \geq 0$, then controller output = [1].

2) If $\sum\limits_{k=0}^{9} OC((n-k)T) > 0$, $|OC(nT)| > 10\%$ controller output range, and $A(nT) < 0$, then controller output = [0].

3) If $\sum\limits_{k=0}^{9} OC((n-k)T) < 0$, $|OC(nT)| > 10\%$ controller output range, and $A(nT) \geq 0$, then controller output = [0].

4) If $\sum\limits_{k=0}^{9} OC((n-k)T) < 0$, $|OC(nT)| > 10\%$ controller output range, and $A(nT) < 0$, then controller output = [-1].

5) Otherwise, If $A(nT) \geq 0$, then controller output = [+]. IF $A(nT) < 0$, then controller output = [-].

Ex. 6.1: *Component Faults*

Figs. 6.8 to 6.10 show respectively the system behaviours caused by different component faults. The fault candidates generated according to the different abnormal behaviours are described together with the figures, where the symbols printed in bold denote the actual faults. The controller stopped tracking and regulating the system when a faulty behaviour was identified. Therefore, the end point of a trace in these figures denotes the time when a fault is detected.

Actual Fault : **R4** blocked at 220 sec.
Observed System State : E11 = [1] (Tank 1 liquid level),
E25 = [1] (Tank 2 liquid level),
E1 = [0] (Controller-1 output),
E15 = [0] (Controller-2 output).
Time for Fault Detection : 8 sec (8 sampling periods).

Diagnosis Result : Power driver fault : Driver-1, Driver-2,
Component block : C2, **R4**.
Measurement Suggestion: F9 (the flow of R2), F23 (the flow of R6),
F6 (rotating speed of Pump-1).

Additional Measurement : F9 = [0], F23 = [0], F6 = [0].
Refined Result : Component block : C2, **R4**.

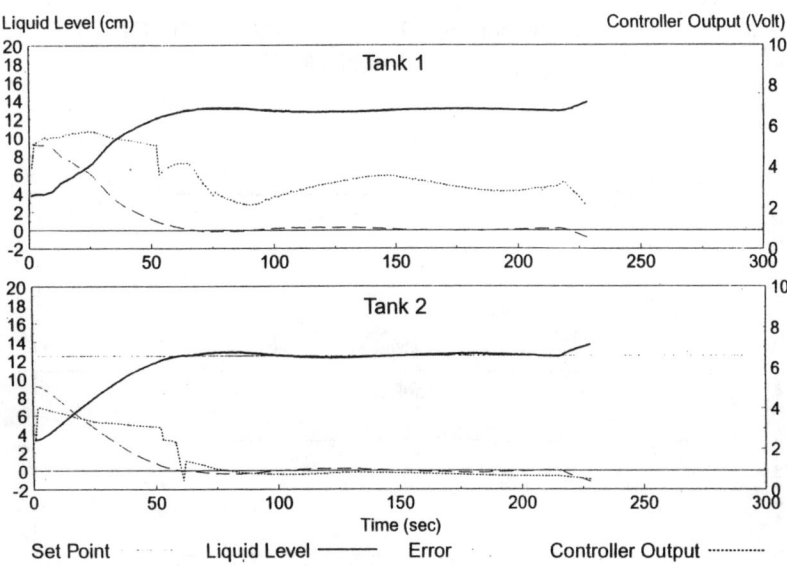

Fig. 6.8 System Behaviour with Tank 2 Discharge Tap Block

Actual Fault   : **C2** leaking at 240 sec.
Observed System State   : E11 = [0], E25 = [0], E1 = [1], E15 = [1].
Time for Fault Detection : 17 sec.

Diagnosis Result   : Power driver fault   : Driver-1, Driver-2,
                            Component leak   : R3, R4,
                            Component leak or block : I1, I2, R1, R2, C1,
                                                     I3, I4, R5, R6, **C2**.
Measurement Suggestion: F9, F23, F6.

Additional Measurement : F9 = [1], F23 = [1], F6 = [1].
Refined Result   : Component leak   : R2, R3, R6, **C2**,
                                                    R4,
                            Component leak or block : C1.
Measurement Suggestion: F13 (the flow of R3), F27 (the flow of R4).

Additional Measurement : F13 = [+], F27 = [0].
Refined Result   : Component leak   : R2, R3, R6, **C2**,
                            Component leak or block : C1.

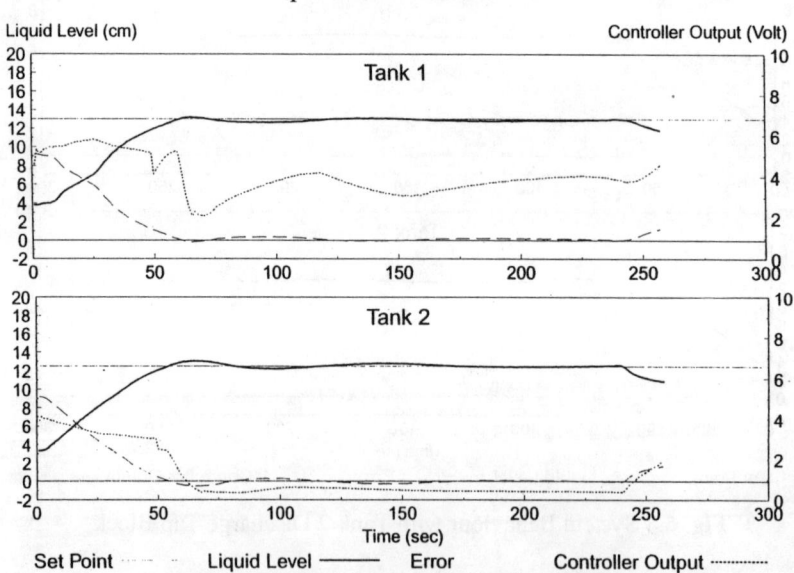

Fig. 6.9 System Behaviour with Tank 2 Leak

Actual Fault   : **R6** leaking at 235 sec.
Observed System State   : E11 = [0], E25 = [0], E1 = [1], E15 = [1].
Time for Fault Detection : 17 sec.

Diagnosis Result   : Power driver fault   : Driver-1, Driver-2,
                             Component leak   : R3, R4,
                             Component leak or block : I1, I2, R1, R2, C1,
                                                        I3, I4, R5, **R6**, C2.

Measurement Suggestion: F9, F23, F6.

Additional Measurement : F9 = [1], F23 = [0], F6 = [1].
Refined Result   : Power driver fault   : Driver-2,
                             Component leak   : R2, R3, C2, R4,
                             Component leak or block : C1, I3, I4, R5, **R6**.

Measurement Suggestion: F13, F27, F20 (rotating speed of Pump-2).

Additional Measurement : F13 = [1], F27 = [0], F20 = [1].
Refined Result   : Component leak   : R2, R3, C2, R5,
                                                           R6,
                             Component leak or block : C1.

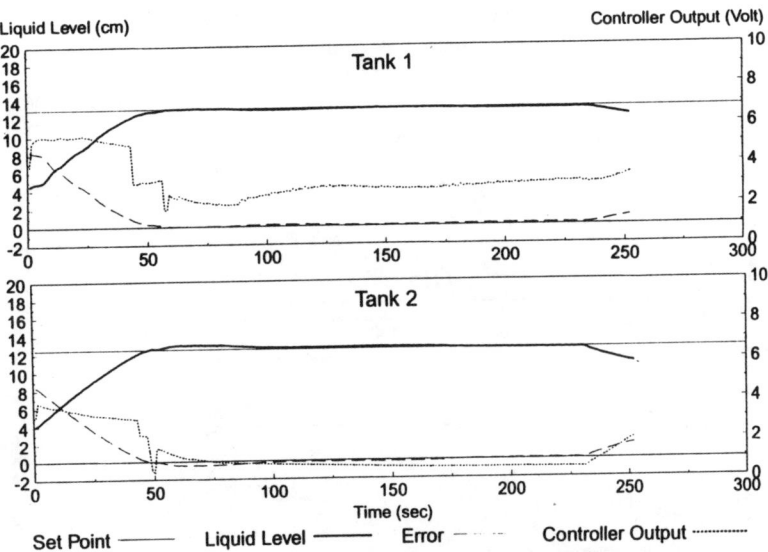

| Set Point ——— | Liquid Level ——— | Error — ·· | Controller Output ············ |

Fig. 6.10 System Behaviour with Tank 2 Inlet Pipe Leak

Ex. 6.2: *Driver Faults*

Figs. 6.11 and 6.12 show the system behaviours caused by driver faults. The fault diagnosis results are shown as follows:

Actual Fault   : **Driver-1** breakdown at 215 sec.
Observed System State   : E11 = [0], E25 = [+], E1 = [1], E15 = [+].
Time for Fault Detection : 11 sec.

Diagnosis Result   : Sensor fault   : Sensor-1, Sensor-2
                                       Power driver fault   : **Driver-1**,
                                       Component leak or block : I1, I2, R1, R2, C1.
Measurement Suggestion: F9, F23, F6.

Additional Measurement : F9 = [0], F23 = [+], F6 = [0].
Refined Result   : Sensor fault: : Sensor-1, Sensor-2
                                       Power driver fault   : **Driver-1**,
                                       Component leak or block : I1, I2, R1.
Measurement Suggestion: F5, F2, F27.

Additional Measurement : F5 = [0], F2 = [0], F27 = [0].
Refined Result   : Sensor fault   : Sensor-1, Sensor-2
                                       Power driver fault   : **Driver-1**,
                                       Component leak or block : I1.

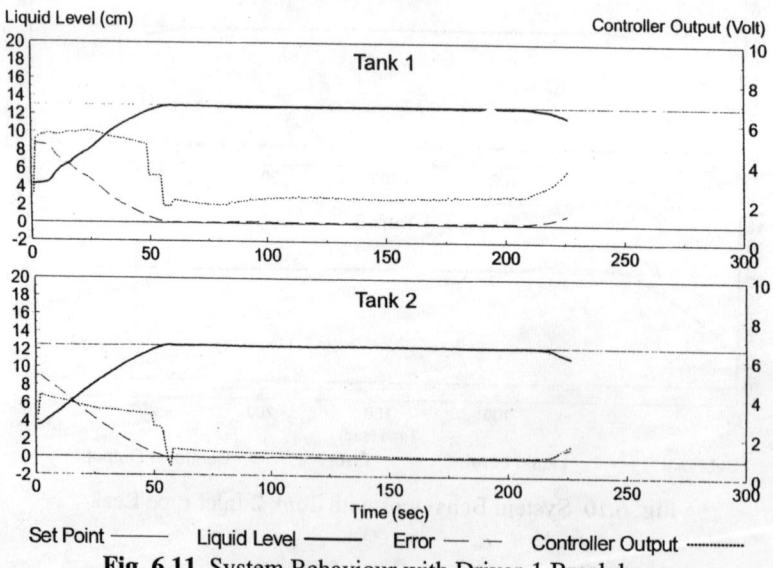

Fig. 6.11 System Behaviour with Driver-1 Breakdown

Actual Fault   : **Driver-2** breakdown at 220 sec.
Observed System State   : E11 = [0], E25 = [0], E1 = [1], E15 = [1].
Time for Fault Detection : 27 sec.

Diagnosis Result   : Power driver fault   : Driver-1, **Driver-2**
                                Component leak   : R3, R4,
                                Component leak or block : I1, I2, R1, R2, C1,
                                                            I3, I4, R5, R6, C2.
Measurement Suggestion: F9, F23, F6.

Additional Measurement : F9 = [1], F23 = [0], F6 = [1].
Refined Result   : Power driver fault   : **Driver-2,**
                                Component leak   : R2, R3, C2, R4,
                                Component leak or block : C1, I3, I4, R5, R6.
Measurement Suggestion: F13, F27, F20.

Additional Measurement : F13 = [1], F27 = [0], F20 = [0].
Refined Result   : Power driver fault   : **Driver-2,**
                                Component leak   : R2, R3, C2,
                                Component leak or block : C1, I3, I4, R5.

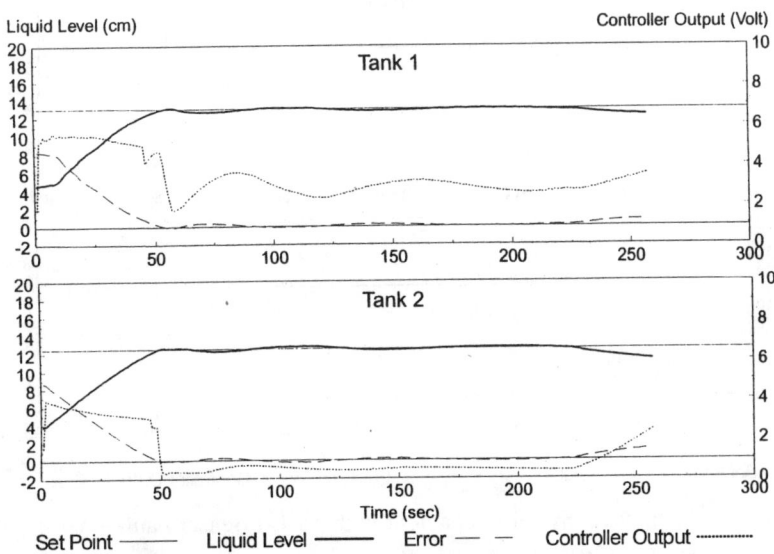

Fig. 6.12 System Behaviour with Driver-2 Breakdown

Ex. 6.3: *Sensor Fault*

Fig. 6.13 shows the system behaviour caused by Sensor-1 fault., and the fault diagnosis result is shown as follows:

Actual Fault   : **Sensor-1** breakdown at 212 sec.
Observed System State : E11 = [0], E25 = [+], E1 = [1], E15 = [+].
Time for Fault Detection : 5 sec.

Diagnosis Result   : Sensor fault   : **Sensor-1**, Sensor-2
                                    Power driver fault   : Driver-1,
                                      Component leak or block : I1, I2, R1, R2, C1.
Measurement Suggestion: F9, F23, F6.

Additional Measurement : F9 = [1], F23 = [+], F6 = [1].
Refined Result   : Sensor fault   : **Sensor-1**, Sensor-2
                                    Component leak   : R2,
                                    Component leak or block : C1.

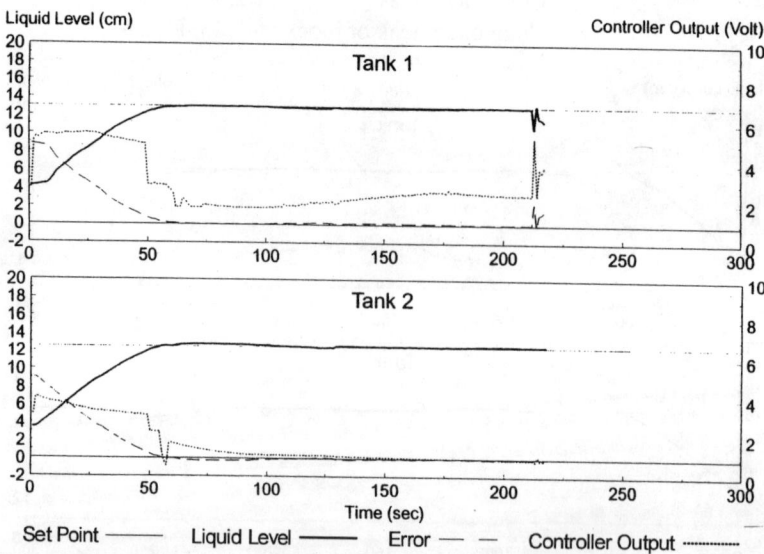

Fig. 6.13 System Behaviour with Tank 1 Sensor Fault

From these experiments, two characteristics of this qualitative fault diagnosis method can be found. One is that the effectiveness of fault localisation is determined at the stage of identifying qualitative values for an abnormal behaviour. For example, the abnormal behaviour identified in the experiment shown in Fig. 6.11 was E11 = [0], E25 = [+], E1 = [1], and E15 = [+]. This identification led to a diagnosis result narrower than the result obtained from the identification of E11 = [0], E25 = [0], E1 = [1], and E15 = [1] (shown in Fig. 6.12). In this fault diagnosis method, an abnormal behaviour is identified according to the performance criteria given by the user. Therefore, setting adequate performance criteria is vital for the effectiveness of this fault diagnosis method. Nevertheless, the setting of performance criteria itself will not affect the correctness of the fault diagnosis result. It can be seen from the experiments shown in Fig. 6.9 to 6.12 that although the initial diagnosis results are different, the actual faults are always included in all the sets of initial fault candidates. That is, the use of inappropriate performance criteria affects only the effectiveness but not the correctness in this fault diagnosis method.

The other characteristic is that, generally, more measurements lead to a more accurate diagnosis result. However, in some circumstances, further measurements cannot very effectively narrow the range of fault candidates. The reason for this is that measurements used in this method are represented in qualitative values and thus cannot be used to distinguish slight differences caused by different faults. Accordingly, some measurements are synonymous for the inference mechanism. For example, in the case shown in Fig. 6.9, the two final additional measurements only eliminated one possible fault. This is because the additional measurement E13 = [+] indicates the state which has already been implied by E11 = [0] and E25 = [0], so that measuring E13 is of no value for refining the diagnosis result. In this circumstance, the user should decide whether more additional measurements are needed or not.

6.8 FEEDBACK COMPONENTS AND THE DETAIL-LEVEL OF MODELS

So far, the cause-effect inference for this fault diagnosis method has been developed based on a premise that system components are independent of power variables. The causal relation which has been considered is that a component parameter change causes power variable changes. However, as addressed by Iwasaki and Simon [1986], the qualitative equations used in this method do not prevent a power variable change from causing a component parameter change. Such a circumstances can happen to feedback components, i.e. as a pressure reducing valve. This section will discuss the cause-effect inference about feedback components and the relationship between the inference result and the detail-level of qualitative bond graph models.

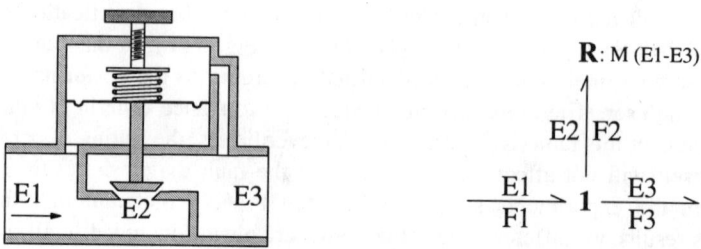

Fig. 6.14 Pressure Reducing Valve and its Simple Bond Graph Model

Fig. 6.14 shows a pressure reducing valve and its simple bond graph model. With the use of the qualitative representation proposed in Chapter 3, the property of this valve is described as:

$$E1(nT) = E2(nT) + E3(nT), \tag{6-45}$$
$$F1(nT) = F2(nT) = F3(nT), \tag{6-46}$$
$$E2(nT) = R \times F2(nT). \tag{6-47}$$

This representation is the same as that for a general valve. Thus, the use of these equations is the same as that of non-feedback components, and the fault diagnosis mechanism can also denote the possible fault (blockage or leakage) for this pressure reducing valve. However, if we want to delve into the original causes which lead to the leakage or blockage of the valve, then this representation appears to be not detailed enough, since it does not describe the function of the valve completely. Thus, a more detailed representation of the pressure reducing valve is required, and it can be written as follows:

$$E1(nT) = E2(nT) + E3(nT), \tag{6-48}$$
$$F1(nT) = F2(nT) = F3(nT), \tag{6-49}$$
$$E2(nT) = R(nT) \times F2(nT), \tag{6-50}$$
$$R(nT) \ = M \times (E1(nT) - E3(nT)),$$
$$\ = M \times E2(nT), \tag{6-51}$$

where M is the modulus of the valve. In this representation, R becomes a variable of $E2$. Through these equations, more causal relations relating to this valve can be inferred. For example, assume that the pressure difference between $E1$ and $E3$ is very big and the mechanism of the valve is normal. Thus, $E1(nT)$ can be given by

[1], $E3(nT)$ can be given by [0], and M can be set as [+]. Then, from Eq. (6-51), $E2(nT)$ can be obtained as [1], and $R(nT)$ can also be obtained as [1]. This means that the valve is blocked. Thus, as discussed in the fault localisation method, $F2(nT)$ will tend to [0], and the flow passing through this valve will also tend to [0]. The relations between system variables are shown as follows:

$$
\begin{aligned}
E1(nT) &= E2(nT) + E3(nT) &\rightarrow& \quad [1] &=& [1] + [0], \\
F1(nT) &= F2(nT) = F3(nT) &\rightarrow& \quad [0] &=& [0] = [0], \\
E2(nT) &= R(nT) \times F2(nT) &\rightarrow& \quad [1] &=& [1] \times [0], \\
R(nT) &= M \times E2(nT) &\rightarrow& \quad [1] &=& [+] \times [1].
\end{aligned}
$$

Here, all the equations are satisfied. That is, a very big pressure difference between $E1$ and $E3$ will cause the valve to be blocked and lead to the flow being zero.

Also, the faults caused by the mechanism problem of the valve can be inferred via the representation. Assume that a mechanism problem causes M to be [1]. Thus, the possible inferences performed according to different pressure inputs are as follows:

1)
$$
\begin{aligned}
E1(nT) &= E2(nT) + E3(nT) &\rightarrow& \quad [1] &=& [1] + [0], \\
F1(nT) &= F2(nT) = F3(nT) &\rightarrow& \quad [0] &=& [0] = [0], \\
E2(nT) &= R(nT) \times F2(nT) &\rightarrow& \quad [1] &=& [1] \times [0], \\
R(nT) &= M \times E2(nT) &\rightarrow& \quad [1] &=& [1] \times [1],
\end{aligned}
$$

2)
$$
\begin{aligned}
E1(nT) &= E2(nT) + E3(nT) &\rightarrow& \quad [+] &=& [+] + [0], \\
F1(nT) &= F2(nT) = F3(nT) &\rightarrow& \quad [0] &=& [0] = [0], \\
E2(nT) &= R(nT) \times F2(nT) &\rightarrow& \quad [+] &=& [1] \times [0], \\
R(nT) &= M \times E2(nT) &\rightarrow& \quad [1] &=& [1] \times [+].
\end{aligned}
$$

3)
$$
\begin{aligned}
E1(nT) &= E2(nT) + E3(nT) &\rightarrow& \quad [0] &=& [0] + [0], \\
F1(nT) &= F2(nT) = F3(nT) &\rightarrow& \quad [0] &=& [0] = [0], \\
E2(nT) &= R(nT) \times F2(nT) &\rightarrow& \quad [0] &=& [?] \times [0], \\
R(nT) &= M \times E2(nT) &\rightarrow& \quad [?] &=& [1] \times [0].
\end{aligned}
$$

All the inferences lead to the result "$F2(nT) = 0$". This means that, in any situation, the mechanism fault $M = [1]$ will cause the flow passing through the valve to tend to be zero.

On the other hand, if a mechanism fault causes M to be [0] and the pressure difference between $E1$ and $E3$ is not very big, then $R(nT)$ can be inferred to be [0]. Thus, $E2(nT)$ will also tend to be [0]. This implies that $E1(nT)$ and $E3(nT)$ will tend to be equal. That is, the mechanism fault causing "$M = [0]$" will lead to the result of "$E1(nT) = E3(nT)$". This inference is shown as:

$$
\begin{aligned}
E1(nT) &= E2(nT) + E3(nT) &\rightarrow\quad [?] &= [0] + [?], \\
F1(nT) &= F2(nT) = F3(nT) &\rightarrow\quad [?] &= [?] = [?], \\
E2(nT) &= R(nT) \times F2(nT) &\rightarrow\quad [0] &= [0] \times [?], \\
R(nT) &= M \times E2(nT) &\rightarrow\quad [0] &= [0] \times [0].
\end{aligned}
$$

Although some ambiguous values appear in this inference, the most important fact of "$E1(nT) = E3(nT)$" can still be confirmed without confusion.

A further problem now arises from the above discussion — what reasons cause the modulus M to be [1] or [0]? To resolve this problem, a more detailed model of the pressure reducing valve is required. This model should involve the detailed information about the spring, the diaphragm, the chambers divided by the diaphragm, and the tubes connecting the chambers to the valve. Thus, more casual relations can be inferred via reasoning about this detailed model. For example, a [1] type fault of the valve can be caused by the blockage of the bottom chamber or its tube, while a [0] type fault can be caused by the broken spring or top chamber. Based on this detailed model, the fault diagnosis mechanism can find the original reasons for the abnormal behaviour.

It can be found from the above discussion that, in the qualitative bond graph reasoning, the depth of the cause-effect inference depends on the detail-level of a qualitative model. A more detailed qualitative model supports reasoning about more detailed causal relations. In the fault diagnosis method, an adequate model-detail-level should be decided by users when considering what the fault diagnosis scheme is used for. From a maintenance standpoint, the simplest model described in Eqs. (6-45) to (6-47) is detailed enough to help find system faults. However, from a design standpoint, the information provided by the simplest model seems to be not detailed enough. For example, if a pressure reducing valve is adjusted to produce a large amount of pressure reduction, then it could be blocked frequently. This circumstance cannot be inferred through the simplest model, but needs a more detailed model (as shown in Eqs. (6-48) to (6-51)). According to the cause-effect inference drawn from the detailed model, designers can adopt some suitable strategies, such as serialising several pressure reducing valves, to avoid frequent system failure.

6.9   DISCUSSION

This chapter has presented a fault diagnosis method based on qualitative bond graph models. These models were generated by the qualitative modelling method proposed in Chapter 3, and they were the same as that used to derive control algorithms for the hybrid qualitative and quantitative controllers in Chapter 4. A set of qualitative descriptors was defined to represent both normal and abnormal behaviours of a physical system. A set of qualitative operations was then defined in terms of the implications of the qualitative descriptors for fault localisation inference. After this, fault candidates of system components, sensors, controllers, and power drivers can be inferred via applying the operations on the qualitative equations derived from the bond graph model of the system. Then, additional measurements can be suggested through reasoning about the system structure described by the qualitative equations to help refine the diagnosis result. The implementation of this method has been illustrated by experiments on a MIMO coupled tanks liquid level control rig in real-time.

Through the implementation, several distinctive advantages of this fault diagnosis method can be summarised as follows:

- Fault diagnosis can be performed without considering system parameters, so that the difficulties of evaluating system parameters correctly can be avoided.

- The qualitative model used in this method is the same as that used for deriving control algorithms for hybrid qualitative and quantitative controllers, and the performance indexes used for fault detection are the same as that employed in the auto-tuning scheme. Therefore, this method is very easy to be combined with a feedback control method.

- This method needs no fault models or fault trees to help infer fault candidates so that the difficulties in building complete fault models can be avoided, and unanticipated faults can be localised.

- The qualitative equations used in this fault diagnosis method contain the complete structural information about a system, so that completeness for fault diagnosis can be guaranteed.

All these benefits are attributed to the properties of qualitative bond graph models and the representation of qualitative equations. Since these equations link all the structural, behavioural, and functional information about a physical system, they are capable of being used for both simulation and inference for fault diagnosis.

Meanwhile, since these equations also represent the locations of system components and their interconnections, the task of fault localisation can thus be extremely simplified. That is, a good "model" is the key to the success of a model-based methodology.

It should be noted that this fault diagnosis method utilises cause-effect inference rather than simulation to generate fault candidates. The reason is that qualitative simulation usually produces all the possible behaviours of a system [Kuipers, 1986], since there are no numerical details to support yielding an explicit solution. Therefore, examining all the possible behaviours to find fault candidates will cause the diagnosis to be inefficient. On the other hand, the cause-effect inference is instigated when the system behaviour has been observed, so it is not necessary to generate all possible behaviours to investigate a fault hypothesis, but it only needs the insertion of the observation and fault hypothesis into qualitative equations to see whether it is reasonable or not. Thus, the diagnosis efficiency can be improved.

Finally, there remain two problems which need further considerations. First, the sensitivity of the fault diagnosis method, which characterises the size of a fault that can be identified, is decided by the setting of performance criteria. System faults can be seen as the component variances which cause the system behaviour to exceed the acceptable range defined by performance criteria. Thus, how to choose appropriate performance indexes for a system with which to set appropriate performance criteria so that normal and abnormal behaviours can be distinguished clearly is the major problem to be considered. For the MIMO coupled tanks apparatus, the simple performance indexes of integral error and integral square error are applicable to indicate abnormal behaviour and the values of performance criteria are easy to be obtained. However, for a more complex system, especially for a system with serious non-linearity, choosing performance indexes and setting their values will be more difficult and require a fuller understanding of the system behaviour. Second, the modelling method and qualitative representation used in this book does not involve consideration about feedback components. However, such components have already been widely used in industry, so extending applications of the modelling and representation methods to feedback components is necessary and will be future work for this approach.

CHAPTER 7

INTELLIGENT SUPERVISORY CONTROL

7.1 INTRODUCTION

An intelligent supervisory control system is required to act like human operators in control rooms to execute the tasks of regulation, monitoring, validation, and decisions making. In such a system, general control knowledge and heuristics for these tasks must be stored and represented in a form applicable for inference engines. Chapter 3 has proposed a qualitative bond graph modelling method, which can be used to build deep-level knowledge models for representing the structural, behavioural, and functional information about complex engineering systems. Based on these deep-level knowledge models, a hybrid qualitative and quantitative control method was developed in Chapter 4 for regulating engineering systems. An auto-tuning scheme was then described in Chapter 5 to validate the hybrid controllers, and a qualitative fault diagnosis method was developed in Chapter 6 for fault detection and localisation.

In this chapter, a management mechanism is developed to combine the above three basic aspects of a supervisory control system. This management mechanism will be employed to choose appropriate strategies, according to the performance monitoring results, to cope with different system situations. A demonstration of this supervisory control system is implemented on the MIMO coupled tanks apparatus used in the previous chapters. Then, the representation of qualitative bond graph models is extended to describe the behaviour of logical switches. An automatic planner based on the extended representation is thus developed to deal with the system start-up and shut-down processes, and the emergency measures. This planner will be used to help operators to make decisions for system operations.

7.2   AN ARCHITECTURE FOR INTELLIGENT SUPERVISORY CONTROL

Architectural approach is an important task in the categories of AI and automatic control, and has been developed since 1970's. A proper architecture makes a machine imitate human intelligence effectively. Intelligent control approaches usually employ a hierarchically intelligent structure which is expected to have autonomy of operation in familiar or unfamiliar environments in response to qualitative commands, with minimum or no operator intervention [Ravindranathan and Leitch, 1994]. For example, Saridis [1983] built an architecture with three levels: organisation, co-ordination and execution (as discussed in Chapter 2), and Rasmussen [1986] constructed a multi level decision making structure with the knowledge level on top. All these architectures determine their structure on a finite set of tasks or functions. Problems are resolved by finding the best sequence of a finite set of tasks in response to a change in the real world situation.

Fig. 7.1 shows the architecture developed for the proposed supervisory control system based on qualitative bond graph models.

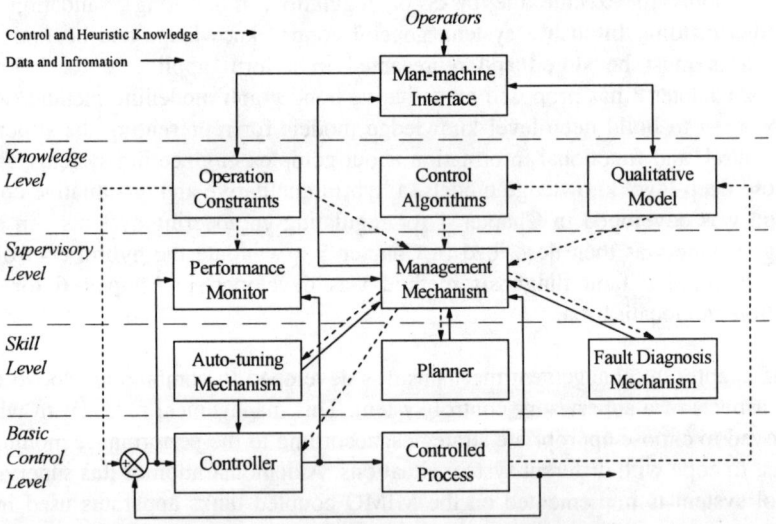

Fig. 7.1 Block Diagram of an Intelligent Supervisory Control System
Based on Qualitative Bond Graph Models

This architecture is composed of four levels: *knowledge level, supervisory level, skill level,* and *basic control level.* Above these levels are human operators who have the responsibility to direct the supervisory system through the man-machine interface. There are several reasons for retaining the human in command. Firstly, the process model has to be built with key variables, such as feedback variables and system input variables, with which the process can be supervised. Human operators are capable of indicating the key variables easily in terms of the control goal.

Secondly, the conditions used to govern the system operation, i.e. set-point and performance criteria, should be provided by operators. This is because the conditions for an engineering system can be changed frequently to produce different products, and human operators are very flexible in setting appropriate conditions for various mission objectives. Besides, some operational constraints which cannot be obtained from logical inferences also should be given by operators. For example, in logical inference, either opening the switch of a motor or braking the rotor can stop the motor. However, we know that braking a motor without opening its switch will damage the motor. In such a situation, operators can specify some rules for a supervisory system to avoid dangerous movements. In the above case, a safety rule can be given by "any operation should not cause the maximum current". On the other hand, "braking a motor without opening its switch will cause the maximum current" can be logically inferred. Thus, the supervisory system can prevent this dangerous movement because the inference result violates the safety rule. Finally, when the result of a fault diagnosis involves operating the process (i.e. opening, or closing a valve), it would be better to let these movements be monitored by human operators. This is because human operators can have more knowledge than a supervisory system to judge whether or not an operation is dangerous.

The *knowledge level* consists of two kinds of knowledge: one is relating to the process structure and the other is the knowledge which cannot be inferred from the process structure. The structural information is obtained via qualitative bond graph modelling and is represented in qualitative equations. The qualitative equations provide a knowledge representation for the fault diagnosis mechanism and the planner to generate cause-effect inference. Also, these equations are used to derive control algorithms for the basic feedback controllers. The detail-level of qualitative models can be decided by users, and known system parameters can be inserted into the qualitative equations to improve the accuracy of the structural information.

On the other hand, the knowledge which cannot be inferred from the qualitative structural information is given by operators. This knowledge includes reference inputs, performance criteria, input and output variables, locations of auxiliary measurements, and operational constraints. The reference inputs are given to the feedback controller, and the performance criteria are sent to the performance monitor to monitor the system behaviour. Other knowledge is assigned to the management mechanism and then allocated to the inference mechanisms to regulate the process.

The *supervisory level* involves the performance monitor and the management mechanism. The performance monitor assesses the system behaviour via comparing the observations of feedback signals, auxiliary measurements, and controller outputs, with the performance criteria. Then, the system behaviour is classified into normal, malfunctioning, or faulty. The observations and comparison results are then passed to the management mechanism. The management mechanism has three functions. Firstly, it allocates the underlying knowledge stored in the knowledge level for inference mechanisms as follows:

Auto-tuning mechanism:	performance criteria.
Fault diagnosis mechanism:	qualitative equations, locations of process inputs, outputs, and auxiliary measurements.
Planner:	qualitative equations, operational constraints, locations of process inputs and outputs.
Controller:	control algorithms.

Secondly, the management mechanism chooses appropriate inference mechanisms for different control problems. If the process needs to be started up or shut down, the planner will be enabled to suggest the sequence of operations. If the performance monitor indicates that the system is normal, the controller is commanded to regulate the process. When the system behaviour is regarded as malfunctioning, the controller will be kept working and the auto-tuning mechanism will be enabled to tune the scaling factors of the controller. When the system behaviour is detected as faulty, the controller will be stopped and the fault diagnosis mechanism will then be commanded to localise system faults. After the fault diagnosis process, the planner will be enabled to suggest emergency measures.

Finally, the management mechanism communicates system states to inference mechanisms and the man-machine interface. When an inference mechanism is enabled, it will ask the management mechanism for the current system states. The information needed for each inference mechanism is:

Auto-tuning mechanism:	values of performance indexes.
Fault diagnosis mechanism:	values of performance indexes, controller outputs.
Planner:	values of performance indexes, controller outputs, fault candidates, switch states.

Also, the inference mechanisms will return their inference results to the management mechanism. All the information, including system states and inference results, are shown to human operators through the man-machine interface.

The *skill level* consists of three inference mechanisms: the auto-tuning mechanism, the fault diagnosis mechanism, and the planner; each one simulating a core skill of human operators. The auto-tuning mechanism copes with the malfunctions caused by process parameter changes, reference input changes, or environment changes. The fault-diagnosis mechanism localises system faults when the process structure changes accidentally, system parameters change outside their designed ranges, or sensors fail. The planner suggests the sequences of start-up, shut-down, and emergency measures to human operators to help them make appropriate decisions.

The *basic control level* contains a hybrid qualitative and quantitative controller, the controlled process, and a feedback loop. The controller works as an ordinary feedback controller and has the basic learning capability to set the initial controller output according to the reference input, which has been discussed in Chapter 4.

7.3 CASE STUDY

In this section, the feedback control, auto-tuning, and fault diagnosis methodologies developed in the earlier chapters are combined together through a management mechanism to supervise the MIMO coupled tanks apparatus. The schematic diagram of this system and the bond graph model of the coupled tanks rig were shown respectively in Figs. 4.22 and 4.23, and the qualitative equations used to represent the model were written in Eqs. (4-52) to (4-91).

There are two operational constraints in the following experiments. First, the fault diagnosis mechanism can be enabled only when steady state is achieved, because the fault diagnosis method is only applicable to the steady state. The steady state is regarded as being reached when $\hat{e}c(nT) < 0$ and $|\hat{e}(nT)| < |Max.\ negative\ \hat{e}c((n-1)nT)|$. Also, if a disturbance is caused by a set-point change, the fault diagnosis mechanism will not be enabled. Second, the error scaling factor SF_e is limited to be not smaller than 0.05. There are two reasons for this. One is that $1/SF_e$ can be seen as the integral constant for the controller, so a very small value of SF_e may cause the system to be unstable. The other is that SF_e is evaluated according to the value of the performance index J_e. A very big J_e will lead to a very small SF_e. However, the very big J_e is usually caused by system faults (i.e. structure changes or sensor failures). Thus, in this situation, tuning the scaling factor cannot modify the abnormal behaviour, but could lead to some undesirable conditions (e.g. instability) of the system.

The performance indexes used here are: J_O (change rate of controller output), $J_{\Delta e}$ (change rate of system error), J_e (integral of system error), and J_{se} (integral of squared error), where J_O and $J_{\Delta e}$ indicate the transient-state performance, while J_e and J_{se} indicate the steady-state performance. In the transient state, the conditions for normal behaviour are: $J_{ol} < J_O < J_{ou}$ and $J_{\Delta el} < |J_{\Delta e}| < J_{\Delta eu}$. In the steady state, the system behaviour is classified into three categories: normal, malfunctioning, and faulty. The malfunctioning behaviour will be modified by the auto-tuning mechanism, while the faulty behaviour will initiate the fault diagnosis mechanism to localise system faults. The conditions for the normal behaviour are: $J_{el} < J_e < J_{eu}$ and $J_{se} < J_{seu}$, the conditions for the faulty behaviour are: $J_e < J_{ef2}$, $J_e > J_{ef1}$, and $J_{se} > J_{sef}$, while the behaviour between these two ranges is seen as malfunctioning behaviour. The performance criteria used here are: $J_{ou1} = 1.04$, $J_{ol1} = 0.96$, $J_{ou2} = 1.04$, $J_{ol2} = 0.96$, $J_{\Delta eu1} = 24.0$, $J_{\Delta el1} = 19.0$, $J_{\Delta eu2} = 24.0$, $J_{\Delta el2} = 19.0$, $J_{eu1} = +300$, $J_{el1} = -300$, $J_{eu2} = +400$, $J_{el2} = -400$, $J_{seu1} = 9000$, $J_{seu2} = 16000$, $J_{ef1} = +1500$, $J_{ef2} = -1500$, and $J_{sef} = 225000$. The methods of setting the performance criteria are the same as in Chapters 5 and 6.

Figs. 7.2 and 7.3 show two experimental results performed by the qualitative supervisory control method. In these experiments, the controller stopped tracking and regulating the system behaviour when a faulty behaviour was identified. Therefore, the end point of a trace in these figures denotes the time when a fault is detected. The symbols printed in bold denote the actual faults unanticipated.

These experiments showed that the management mechanism can supervise the coupled tanks apparatus well. A problem rising from the combination of auto-tuning and fault diagnosis mechanisms is that it prolongs the time for fault detection. It can be seen from Fig. 7.2 that the behaviour caused by the blockage of R4 was firstly identified as a malfunctioning behaviour. The scaling factor SF_{e2} was tuned to 0.12 at 255 sec to adapt the system change. However, this tuning was not capable of modifying the behaviour caused by the system structural change, so the system error kept increasing fast. Then, this behaviour was identified as a fault at 263 sec. In comparing this result to that of Ex. 6.1 the same fault was identified 8 seconds after it happened, while in this experiment, it took 13 seconds. This is because the supervisory system spent a few seconds trying to adapt to the system fault. On the other hand, in the experiment shown in Fig. 7.3, the sensor failure caused a very big jump in the system output of Tank 1 liquid level, so this behaviour could be identified as a fault directly without time delay. In this supervisory system, the time

Actual Fault   : **R4** blocked at 250 sec.
Observed System State   : E11 = [1], E25 = [1], E1 = [0], E15 = [0].
Time for Fault Detection : 13 sec (13 sampling periods).

Diagnosis Result   : Power driver fault : Driver-1, Driver-2,
                              Component block   : C2, **R4**.
Measurement Suggestion: F9, F23, F6.

Additional Measurement : F9 = [0], F23 = [0], F6 = [0].
Refined Result   : Component block : C2, **R4**.

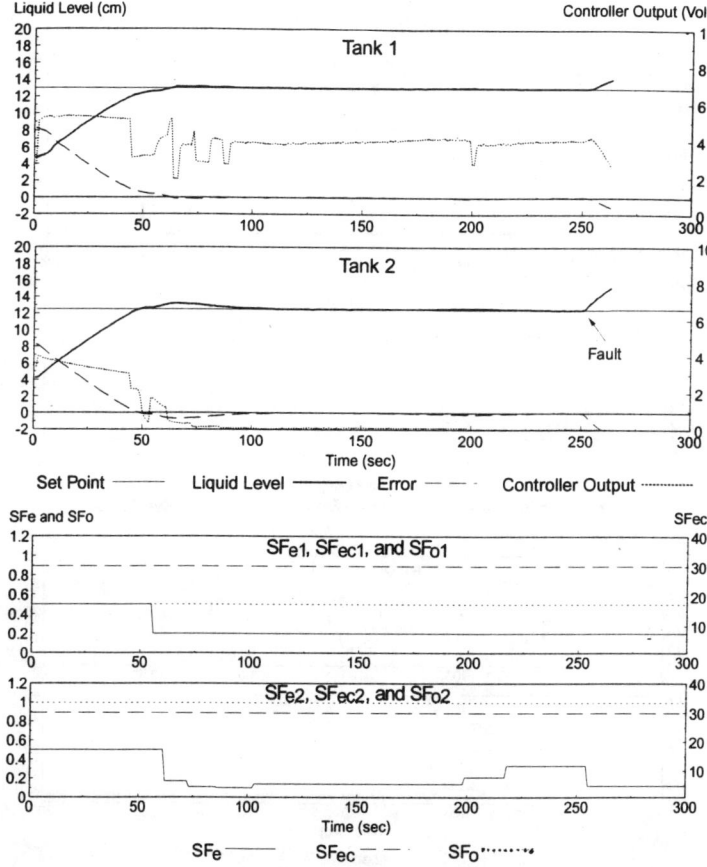

Fig. 7.2 System Behaviour with Tank 2 Discharge Tap Block

Actual Fault   : **Sensor-1** breakdown at 252 sec.
Observed System State   : E11 = [0], E25 = [+], E1 = [1], E15 = [+].
Time for Fault Detection : 1 sec.
Diagnosis Result   : Sensor fault   : **Sensor-1**, Sensor-2
                                        Power driver fault   : Driver-1,
                                        Component leak or block : I1, I2, R1, R2, C1.
Measurement Suggestion: F9, F23, F6.
Additional Measurement : F9 = [1], F23 = [+], F6 = [1].
Refined Result   : Sensor fault   : **Sensor-1**, Sensor-2
                                        Component leak   : R2,
                                        Component leak or block : C1.

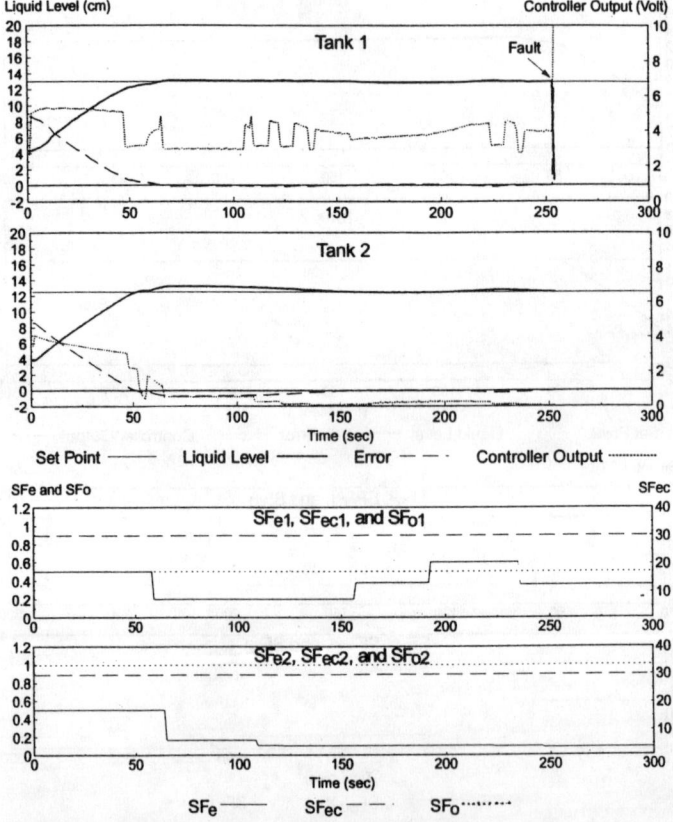

Fig. 7.3 System Behaviour with Tank 1 Sensor Fault

for fault detection can be reduced by using smaller J_{ef1} and J_{ef2} or performance indexes which can distinguish the malfunctioning behaviour and faulty behaviour more effectively.

7.4   QUALITATIVE REPRESENTATION FOR SWITCHES

The qualitative bond graph modelling and representation methods developed in the previous chapters were focused on continuous components. However, a real engineering system usually contains a number of discontinuous components, such as switches and sequence valves, for interlocking a sequence. A supervisory control system should be capable of dealing with both continuous and discontinuous components. Therefore, the schemes of qualitative bond graph modelling and representation are extended here to cope with discontinuous components so that the supervisory control method can be applied to real engineering systems. The bond graph representation of discontinuous components was explored by Lorenz [1993]. However, no constitutive law has been given to describe the switches' behaviour by formal equations. In the 1995 International Conference on Bond Graph Modelling and Simulation [Cellier and Granda, 1995], a variety of approaches have been proposed to try to find a general bond graph description for discontinuous components. In this section, the representation of qualitative equations is employed to represent the constitutive law of switches, and the qualitative descriptors proposed in Chapter 6 are used to describe the states of switches.

The bond graph symbol of a switch is written as

$$\longrightarrow S,$$

and its constitutive equation is

$$E = S \times F. \tag{7-1}$$

Since switches do not store and release energy, the half arrow on the bond of a switch always points towards the switch. A switch can be connected to a system by a 1-junction or 0-junction. The modelling method of switches is the same as that for resistance (see Section 3.4.2). The parameter S represents the states of a switch. The value of S is discrete rather than continuous. The qualitative descriptors [1] and [0] proposed in Chapter 6 can be used to represent the states of switches. The value [1] describes the disconnecting state for a component, i.e. an opened switch or a closed valve, where the flow passing through the component is zero. The value [0] describes the connecting state for a component, i.e. a closed switch or a opened valve, where the effort working on the component is zero. With the use of the qualitative descriptors, the behaviour of a switch can be represented, and the cause-effect inference can be performed through the qualitative operations proposed in Chapter 6. The following example shows how to infer the behaviour of a discontinuous system.

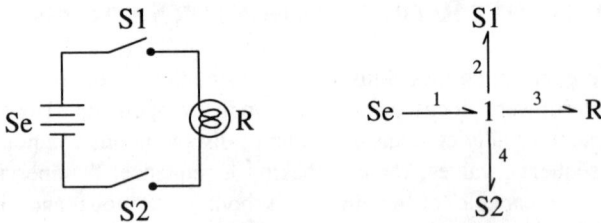

Fig. 7.4 Serial Switches and the Bond Graph Model

Fig. 7.4 shows the structure of a simple electrical circuit with two switches connected serially together with its bond graph model. The qualitative equations of this circuit are

$$E1(nT) = E2(nT) + E3(nT) + E4(nT),\qquad\qquad (7\text{-}2)$$
$$F1(nT) = F2(nT) = F3(nT) = F4(nT),\qquad\qquad (7\text{-}3)$$
$$E2(nT) = S1 \times F2(nT),\qquad\qquad (7\text{-}4)$$
$$E3(nT) = R \times F3(nT),\qquad\qquad (7\text{-}5)$$
$$E4(nT) = S2 \times F4(nT).\qquad\qquad (7\text{-}6)$$

Let us assume that the power source and the lamp are normal, S1 is closed, and S2 is opened. Thus, the system behaviour can be inferred through the following operations:

$$E1(nT) = E2(nT) + E3(nT) + E4(nT) \quad \rightarrow \quad [+] = [0] + [0] + [+],$$
$$F1(nT) = F2(nT) = F3(nT) = F4(nT) \quad \rightarrow \quad [0] = [0] = [0] = [0],$$
$$E2(nT) = S1 \times F2(nT) \quad \rightarrow \quad [0] = [0] \times [0],$$
$$E3(nT) = R \times F3(nT) \quad \rightarrow \quad [0] = [+] \times [0],$$
$$E4(nT) = S2 \times F4(nT) \quad \rightarrow \quad [+] = [1] \times [0].$$

It can be seen that, when S2 is opened, the current in this circuit will be zero and the potential of S2 will equal that of the power source. Since the current flowing through the lamp is zero, the lamp will not work. Similarly, when S1 is opened or both of the switches are opened, the current in this circuit will be zero and the lamp will not work. It can be inferred that the lamp will light only when both the switches are closed. This inference is written as follows:

$$E1(nT) = E2(nT) + E3(nT) + E4(nT) \quad \rightarrow \quad [+] = [0] + [+] + [0],$$
$$F1(nT) = F2(nT) = F3(nT) = F4(nT) \quad \rightarrow \quad [+] = [+] = [+] = [+],$$
$$E2(nT) = S1 \times F2(nT) \quad \rightarrow \quad [0] = [0] \times [+],$$
$$E3(nT) = R \times F3(nT) \quad \rightarrow \quad [+] = [+] \times [+],$$
$$E4(nT) = S2 \times F4(nT) \quad \rightarrow \quad [0] = [0] \times [+].$$

In this situation, the voltage decay of the lamp is equal to the voltage supplied by the power source $(E1(nT) = E3(nT))$.

It can be seen from this example that switches can be modelled and represented easily in conjunction with continuous components. The qualitative descriptors and operations defined in Chapter 6 can be used to infer the behaviour of systems involving switches. Based on these qualitative operations, the following section will develop a series of inference procedures to suggest operation sequences for the qualitative supervisory control system.

7.5   DERIVATION OF OPERATION SEQUENCES

This section develops three generic inference procedures to suggest the operation sequences for system start-up, shut-down, and emergency measures. The basic idea for these inference methods is similar to that of the cause-effect inference method used for fault diagnosis. Firstly, the inference mechanism inserts the current system states and states which are expected to be reached into the qualitative equations of the system. Secondly, the inference mechanism inserts the possible states of switches into the equations, and then induces all unknown variables in the equations. Thirdly, the recognisability of the equations is investigated to find the applicable switch states. Finally, all the variables in the recognised equations are compared to the operational constraints given by users to find the best sequence of operation for the switches.

The detailed inference procedures are given as follows:

- *Deriving the Start-up Sequence:*

    1)   Assume all the components in a system to be normal and all the switches to be disconnected.

2) Set the past states $(E((n-1)T)$ or $F((n-1)T))$ of variables of C and I components as [0]. The value [0] is the initial state of the process. Then, set the current states $(E(nT)$ or $F(nT))$ of output variables as [+] (when set-points are positive) or [-] (when set-points are negative). The value [+] or [-] can be seen as the expected state for an output variable.

3) Search the power paths which interlock the system input variables and output variables.

4) Assume all the switches in a power path to be connected.

5) Insert the states of all switches into the equations, and then infer the values of all the unknown variables. Then, investigate whether or not the equations are satisfied.

6) If all the equations are satisfied, then the assumption about the switch states is accepted. If any equation is violated, then assume that the switches in another power path are connected and repeat Steps 4) to 6) until all equations are satisfied. So far, the switches which should be connected have been found. The following steps are used to decide the sequence for connecting these switches.

7) Store the locations of the switches which should be connected and then reset those switches to be disconnected. Next, if the value of the input variable obtained from Step 5) is [+], then reset the value as [1], while, if the value of the input variable is obtained as [-], then reset it as [-1]. The reason for this setting is that the disconnected switches obstruct the power delivery, so the value of the output variable cannot reach the set-point and the feedback controller will increase the system input to the maximum value ([1] or [-1]).

8) Assume that the switch nearest to the input variable is now connected.

9) Infer the value of all the unknown variables and compare the inference results with operational constraints. If no state variable violates the constraints, then the component can be connected first. Then, assume that the switch nearest to this switch is connected, and repeat this step. On the other hand, if any state variable violates the constraints, then find the switch which has violated a variable and assume it to be connected. Then, repeat this step until all the switches which should be connected have been connected.

10)   Reiterate Steps 4) to 9) until all the power paths have been investigated.

- *Deriving the Shut-down Sequence:*

   1)   Assume all the components to be normal. Then, set the output variables and auxiliary measurements as [0] which is the expected state.

   2)   If a controller output (system input) is positive, then set the related input variable as [1]. If a controller output is negative, then set the related input variable as [-1]. The reason for this is the same as Step 7) of the start-up sequence derivation process.

   3)   Search a power path where the switches are connected. These switches need to be disconnected.

   4)   Assume that the connected switch nearest to the input variable is disconnected.

   5)   Infer the values of all the unknown variables, and then compare the inference results to the operational constraints. If no variable violates the operational constraints, then the switch can be disconnected first. Find the switch nearest to this switch, and assume it to be disconnected. Then repeat this step. On the other hand, if any variable violates the operational constraints, then the switch which has a violated variable should be disconnected. Then, repeat this step.

   6)   Reiterate Steps 3) to 5) until all the power paths have been disconnected.

- *Deriving the Emergency Measures Sequence:*

   1)   Insert the controller outputs and auxiliary measurements into the qualitative equations. If the observation of a feedback measurement is [1] or [-1], then set the measured variable as [0]. If the observation is [0], then set the measured variable as [1] (when its set-point is positive) or [-1] (when its set-point is negative). If the observation is [+] or [-], then set the measured variable as the observed value.

   2)   Infer the qualitative values of all unknown variables. If all the qualitative equations are satisfied, then the switch states in the power paths which interlock system inputs and outputs are acceptable and do not need to be

changed. Then go to Step 4). If any of the equations are violated, then go to Step 3).

3)   If fault candidates involve controllers, power drivers, or sensors, then they should be replaced by their spares. Then, go to Step 4). If the replacement must be performed when the system is stopped, then generate the shut-down sequence. On the other hand, if fault candidates do not involve controllers, power drivers, or sensors, then the switches in the power paths which contain the violated variables obtained from Step 2) should be disconnected. Then, go to Step 4).

4)   Search for the power paths which contain disconnected switches and connect to the paths that contain fault candidates. Then, assume that the disconnected switches on such a power path are now connected and insert this assumption into the qualitative equations. Next, infer all the qualitative values of unknown variables. If all the equations are satisfied, then these switches should be connected.

5)   So far, the switches which should be connected or disconnected have been found. Thus, reiterate Step 7) and 9) of the start-up sequence derivation procedure to generate the sequence for connecting the switches, and reiterate Step 4) and 5) of the shut-down sequence derivation procedure to generate the sequence for disconnecting the switches.

The method of inferring the values for unknown variables is the same as that used in fault diagnosis and has been discussed in Section 6.5.

7.6   AN EXAMPLE OF OPERATION SEQUENCE DERIVATION

Fig. 7.5 shows a schematic diagram of a coupled tanks apparatus in which an extra motor and pump set is added to the system as a spare part. The use of motors and pumps is operated by switching S1, S2, S3, and S4, where S1 and S3 are electrical switches, and S2 and S4 are directional control valves. Another directional control valve S5 is employed to release the pressure exceeding its limit in the tank C2. The sensors PIC and FI are used to measure liquid level in C2 and the flow rate in R2 respectively. The control goal is to regulate the liquid level in C2. The bond graph model of this apparatus is shown in Fig. 7.6.

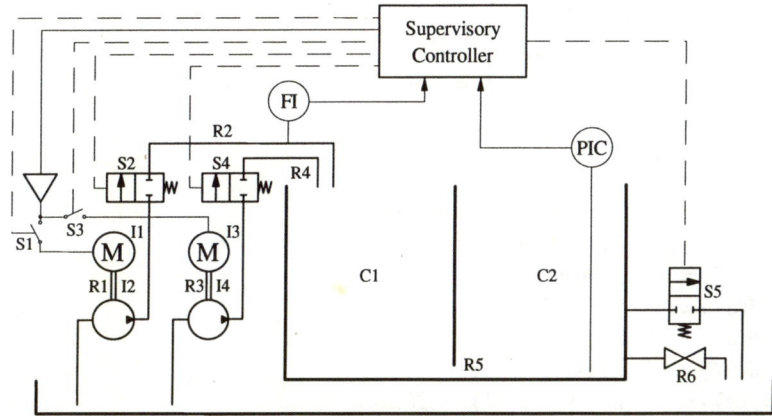

Fig. 7. Schematic Diagram of a Coupled Tanks Liquid Level Control Rig

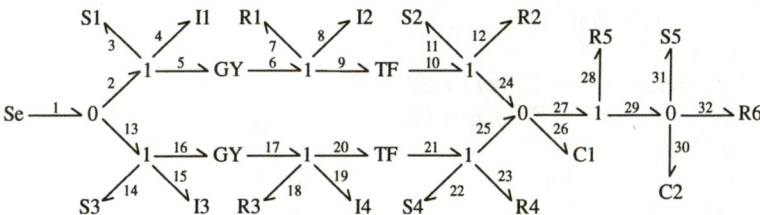

Fig. 7.6 The Bond Graph Model of the Coupled Tanks Apparatus

According to the bond graph model, a set of qualitative equations which represent the property of this apparatus are derived as follows:

$$
\begin{aligned}
E1(nT) &= E2(nT) = E13(nT), & (7\text{-}7)\\
F1(nT) &= F2(nT) + F13(nT), & (7\text{-}8)\\
E2(nT) &= E3(nT) + E4(nT) + E5(nT), & (7\text{-}9)\\
F2(nT) &= F3(nT) = F4(nT) = F5(nT), & (7\text{-}10)\\
E3(nT) &= S1 \times F3(nT), & (7\text{-}11)\\
E4(nT) &= I1 \times (F4(nT) - F4(n\text{-}1)T)), & (7\text{-}12)\\
E5(nT) &= F6(nT), & (7\text{-}13)\\
F5(nT) &= E6(nT), & (7\text{-}14)\\
E6(nT) &= E7(nT) + E8(nT) + E9(nT), & (7\text{-}15)\\
F6(nT) &= F7(nT) = F8(nT) = F9(nT), & (7\text{-}16)\\
E7(nT) &= R1 \times F7(nT), & (7\text{-}17)\\
E8(nT) &= I2 \times (F8(nT) - F8((n\text{-}1)T)), & (7\text{-}18)\\
E9(nT) &= E10(nT), & (7\text{-}19)
\end{aligned}
$$

$$F9(nT) = F10(nT), \tag{7-20}$$
$$E10(nT) = E11(nT) + E12(nT) + E24(nT), \tag{7-21}$$
$$F10(nT) = F11(nT) = F12(nT) = F24(nT), \tag{7-22}$$
$$E11(nT) = S2 \times F11(nT), \tag{7-23}$$
$$E12(nT) = R2 \times F12(nT), \tag{7-24}$$
$$E24(nT) = 0, \tag{7-25}$$
$$E13(nT) = E14(nT) + E15(nT) + E16(nT), \tag{7-26}$$
$$F13(nT) = F14(nT) = F15(nT) = F16(nT), \tag{7-27}$$
$$E14(nT) = S3 \times F14(nT), \tag{7-28}$$
$$E15(nT) = I3 \times (F15(nT) - F15(n\text{-}1)T)), \tag{7-29}$$
$$E16(nT) = F17(nT), \tag{7-30}$$
$$F16(nT) = E17(nT), \tag{7-31}$$
$$E17(nT) = E18(nT) + E19(nT) + E20(nT), \tag{7-32}$$
$$F17(nT) = F18(nT) = F19(nT) = F20(nT), \tag{7-33}$$
$$E18(nT) = R3 \times F18(nT), \tag{7-34}$$
$$E19(nT) = I4 \times (F19(nT) - F19((n\text{-}1)T)), \tag{7-35}$$
$$E20(nT) = E21(nT), \tag{7-36}$$
$$F20(nT) = F21(nT), \tag{7-37}$$
$$E21(nT) = E22(nT) + E23(nT) + E25(nT), \tag{7-38}$$
$$F21(nT) = F22(nT) = F23(nT) = F25(nT), \tag{7-39}$$
$$E22(nT) = S4 \times F22(nT), \tag{7-40}$$
$$E23(nT) = R4 \times F23(nT), \tag{7-41}$$
$$E25(nT) = 0, \tag{7-42}$$
$$E26(nT) = E27(nT), \tag{7-43}$$
$$F24(nT) + F25(nT) = F26(nT) + F27(nT), \tag{7-44}$$
$$F26(nT) = C1 \times (E26(nT) - E26((n\text{-}1)T)), \tag{7-45}$$
$$E27(nT) = E28(nT) + E29(nT), \tag{7-46}$$
$$F27(nT) = F28(nT) = F29(nT), \tag{7-47}$$
$$E28(nT) = R5 \times F28(nT), \tag{7-48}$$
$$E29(nT) = E30(nT) = E31(nT) = E32(nT), \tag{7-49}$$
$$F29(nT) = F30(nT) + F31(nT) + F32(nT), \tag{7-50}$$
$$F30(nT) = C2 \times (E30(nT) - E30((n\text{-}1)T)), \tag{7-52}$$
$$E31(nT) = S5 \times F31(nT), \tag{7-51}$$
$$E32(nT) = R6 \times F32(nT). \tag{7-53}$$

7.6.1   The Start-up Sequence

To operate a supervisory control system, the user should first indicate the key variables of a process. For example, in this case, $E30$ is the output variable, $F12$ is the auxiliary measurement, and $E1$ is the input variable. Also, the user should give the set-point and the operational constraints. In this electro-hydraulic case, the constraint is given that any operation should not result in the maximum pressure for the piping system. In other words, the effort variables of R1, R4, S2, and S4 should never be [1]. After this, the start-up sequence can be derived through the following inference:

Firstly, the initial states of components C and I are given by [0] and $E30(nT)$ is given by [+] (since the set-point here is always positive). Then, according to the structural information provided by the qualitative equations, the power paths which interlock the input and feedback variables can be found. For example, $E30$ is known as the output variable and located at Bond-30. From Eqs. (7-49) and (7-50), it can be found that the power of Bond-30 comes from Bond-29. Then, from Eqs. (7-46) and (7-47), the power of Bond-29 comes from Bond-27. This search will continue until the output variable is linked to the input variable. Through this searching process, two power paths are obtained as: Bonds-1, 2, 5, 6, 9, 10, 24, 27, 29, 30, and Bonds-1, 13, 16, 17, 20, 21, 25, 27, 29, 30. Next, the switches on the former power path (S1 and S2) are assumed to be connected and the switch states (S1 = [0], S2 = [0], S3 = [1], S4 = [1]) can be inserted into the qualitative equations. Thus, all the values in the qualitative equations can be inferred as follows. The serial numbers written in front of the equations indicate the inference sequence. Firstly, the unknown variables in the equations numbered 1) can be obtained. With these newly obtained values, the unknown variables in the equations numbered 2) can then be evaluated. Similarly, other unknown variables can be obtained in order of the sequence number.

19) $E1(nT) = E2(nT) = E13(nT)$ $\rightarrow$ $[+] = [+] = [+],$

17) $F1(nT) = F2(nT) + F13(nT)$ $\rightarrow$ $[+] = [+] + [0],$

18) $E2(nT) = E3(nT) + E4(nT) + E5(nT)$ $\rightarrow$ $[+] = [0] + [+] + [+],$

16) $F2(nT) = F3(nT) = F4(nT) = F5(nT)$ $\rightarrow$ $[+] = [+] = [+] = [+],$

17) $E3(nT) = S1 \times F3(nT)$ $\rightarrow$ $[0] = [0] \times [+],$

17) $E4(nT) = I1 \times (F4(nT) - F4(n-1)T))$ $\rightarrow$ $[+] = [+] \times ([+] - [0]),$

13) $E5(nT) = F6(nT)$ $\rightarrow$ $[+] = [+],$

15) $F5(nT) = E6(nT)$ $\rightarrow$ $[+] = [+],$

14) $E6(nT) = E7(nT) + E8(nT) + E9(nT)$ $\rightarrow$ $[+] = [+] + [+] + [+],$

12) $F6(nT) = F7(nT) = F8(nT) = F9(nT)$ $\rightarrow$ $[+] = [+] = [+] = [+],$

13) $E7(nT) = R1 \times F7(nT)$ $\rightarrow$ $[+] = [+] \times [+],$

13) $E8(nT) = I2 \times (F8(nT) - F8((n-1)T))$ $\rightarrow$ $[+] = [+] \times ([+] - [0]),$

13) $E9(nT) = E10(nT)$ $\rightarrow$ $[+] = [+],$

11) $F9(nT) = F10(nT)$ $\rightarrow$ $[+] = [+],$

12) $E10(nT) = E11(nT) + E12(nT) + E24(nT)$ $\rightarrow$ $[+] = [0] + [+] + [0],$

10) $F10(nT) = F11(nT) = F12(nT) = F24(nT)$ $\rightarrow$ $[+] = [+] = [+] = [+],$

11) $E11(nT) = S2 \times F11(nT)$ $\rightarrow$ $[0] = [0] \times [+],$

11) $E12(nT) = R2 \times F12(nT)$ $\rightarrow$ $[+] = [+] \times [+]$,

1) $E24(nT) = 0$ $\rightarrow$ $[0] = [0]$,

20) $E13(nT) = E14(nT) + E15(nT) + E16(nT)$ $\rightarrow$ $[+] = [+] + [0] + [0]$,

1) $F13(nT) = F14(nT) = F15(nT) = F16(nT)$ $\rightarrow$ $[0] = [0] = [0] = [0]$,

21) $E14(nT) = S3 \times F14(nT)$ $\rightarrow$ $[+] = [1] \times [0]$,

2) $E15(nT) = I3 \times (F15(nT) - F15(n-1)T))$ $\rightarrow$ $[0] = [+] \times ([0] - [0])$,

4) $E16(nT) = F17(nT)$ $\rightarrow$ $[0] = [0]$,

2) $F16(nT) = E17(nT)$ $\rightarrow$ $[0] = [0]$,

5) $E17(nT) = E18(nT) + E19(nT) + E20(nT)$ $\rightarrow$ $[0] = [0] + [0] + [0]$,

3) $F17(nT) = F18(nT) = F19(nT) = F20(nT)$ $\rightarrow$ $[0] = [0] = [0] = [0]$,

4) $E18(nT) = R3 \times F18(nT)$ $\rightarrow$ $[0] = [+] \times [0]$,

4) $E19(nT) = I4 \times (F19(nT) - F19((n-1)T))$ $\rightarrow$ $[0] = [+] \times ([0] - [0])$,

6) $E20(nT) = E21(nT)$ $\rightarrow$ $[0] = [0]$,

2) $F20(nT) = F21(nT)$ $\rightarrow$ $[0] = [0]$,

7) $E21(nT) = E22(nT) + E23(nT) + E25(nT)$ $\rightarrow$ $[0] = [0] + [0] + [0]$,

1) $F21(nT) = F22(nT) = F23(nT) = F25(nT)$ $\rightarrow$ $[0] = [0] = [0] = [0]$,

8) $E22(nT) = S4 \times F22(nT)$ $\rightarrow$ $[0] = [1] \times [0]$,

2) $E23(nT) = R4 \times F23(nT)$ $\rightarrow$ $[0] = [+] \times [0]$,

1) $E25(nT) = 0$ $\rightarrow$ $[0] = [0]$,

7) $E26(nT) = E27(nT)$ $\rightarrow$ $[+] = [+]$,

9) $F24(nT) + F25(nT) = F26(nT) + F27(nT)$ $\rightarrow$ $[+] + [0] = [+] + [+]$,

8) $F26(nT) = C1 \times (E26(nT) - E26((n-1)T))$ $\rightarrow$ $[+] = [+] \times ([+] - [0])$,

6) $E27(nT) = E28(nT) + E29(nT)$ $\rightarrow$ $[+] = [+] + [+]$,

4) $F27(nT) = F28(nT) = F29(nT)$ $\rightarrow$ $[+] = [+] = [+]$,

5) $E28(nT) = R5 \times F28(nT)$ $\rightarrow$ $[+] = [+] \times [+]$,

1) $E29(nT) = E30(nT) = E31(nT) = E32(nT)$ $\rightarrow$ $[+] = [+] = [+] = [+]$,

3) $F29(nT) = F30(nT) + F31(nT) + F32(nT)$ $\rightarrow$ $[+] = [+] + [0] + [+]$,

1) $F30(nT) = C2 \times (E30(nT) - E30((n-1)T))$ $\rightarrow$ $[+] = [+] \times ([+] - [0])$,

2) $E31(nT) = S5 \times F31(nT)$ $\rightarrow$ $[+] = [1] \times [0]$,

2) $E32(nT) = R6 \times F32(nT)$ $\rightarrow$ $[+] = [+] \times [+]$.

It can be seen that all the above equations are satisfied. This means that the control goal can be achieved when S1 and S2 are connected. The next step is to decide the sequence for connecting S1 and S2. For this object, S1 and S2 are again assumed to be disconnected. Meanwhile, since the value of the input variable $E1(nT)$ is obtained from the above inference as $[+]$, $E1(nT)$ is reset as $[1]$. Then, the switch nearest to the input variable is assumed to be connected, and the unknown variables can be evaluated. A part of the inference is shown as follows:

1) $E1(nT)$ $= E2(nT) = E13(nT)$ $\rightarrow$ $[1] = [1] = [1]$,
8) $F1(nT)$ $= F2(nT) + F13(nT)$ $\rightarrow$ $[1] = [1] + [0]$,
5) $E2(nT)$ $= E3(nT) + E4(nT) + E5(nT)$ $\rightarrow$ $[1] = [0] + [1] + [0]$,
7) $F2(nT)$ $= F3(nT) = F4(nT) = F5(nT)$ $\rightarrow$ $[1] = [1] = [1] = [1]$,
8) $E3(nT)$ $= S1 \times F3(nT)$ $\rightarrow$ $[0] = [0] \times [1]$,
6) $E4(nT)$ $= I1 \times (F4(nT) - F4(n-1)T))$ $\rightarrow$ $[1] = [+] \times ([1] - [0])$,
4) $E5(nT)$ $= F6(nT)$ $\rightarrow$ $[0] = [0]$,
8) $F5(nT)$ $= E6(nT)$ $\rightarrow$ $[1] = [1]$,
9) $E6(nT)$ $= E7(nT) + E8(nT) + E9(nT)$ $\rightarrow$ $[1] = [0] + [0] + [1]$,
3) $F6(nT)$ $= F7(nT) = F8(nT) = F9(nT)$ $\rightarrow$ $[0] = [0] = [0] = [0]$,
4) $E7(nT)$ $= R1 \times F7(nT)$ $\rightarrow$ $[0] = [+] \times [0]$,
4) $E8(nT)$ $= I2 \times (F8(nT) - F8((n-1)T))$ $\rightarrow$ $[0] = [+] \times ([0] - [0])$,
10) $E9(nT)$ $= E10(nT)$ $\rightarrow$ $[1] = [1]$,
2) $F9(nT)$ $= F10(nT)$ $\rightarrow$ $[0] = [0]$,
11) $E10(nT) = E11(nT) + E12(nT) + E24(nT)$ $\rightarrow$ $[1] = [1] + [0] + [0]$,
1) $F10(nT) = F11(nT) = F12(nT) = F24(nT)$ $\rightarrow$ $[0] = [0] = [0] = [0]$,
12) $E11(nT) = S2 \times F11(nT)$ $\rightarrow$ $[1] = [1] \times [0]$,
2) $E12(nT) = R2 \times F12(nT)$ $\rightarrow$ $[0] = [+] \times [0]$,
1) $E24(nT) = 0$ $\rightarrow$ $[0] = [0]$,

It can be seen from this inference result that the value of the effort variable of S2 ($E11(nT)$) is [1], which violates the given operational constraints. This means that connecting S1 first is not acceptable for this system. Then, S2 which contains a violated variable is considered to be connected. As discussed in Section 7.4, the effort of a connected switch will be zero. In other words, the value of $E11(nT)$ will be [0] and will not violate the operational constraint. The start-up sequence is thus obtained as: connecting S2 first and connecting S1 afterwards (S2 $\rightarrow$ S1). Similarly, connecting S3 and S4 also makes the system work, and the start-up sequence is "S4 $\rightarrow$ S3". The user can now decide how to operate these switches.

7.6.2   The Shut-down Sequence

Now, suppose that the system will be shut down in a situation in which S1 and S2 are connected whilst S3 and S4 are disconnected. First, all the components are assumed to be normal, and the output variable ($E30(nT)$) and the auxiliary measurement ($F12(nT)$) are set as [0]. The input variable $E1(nT)$ is set as [1], since the normal system input for this system is positive. Then, switch S1, which is the one nearest the system input, is assumed to be disconnected first. Thus, the values of unknown variables can be inferred. A part of this inference is shown as follows:

1) $E1(nT) = E2(nT) = E13(nT)$ $\rightarrow$ $[1] = [1] = [1],$
2) $F1(nT) = F2(nT) + F13(nT)$ $\rightarrow$ $[0] = [0] + [0],$
3) $E2(nT) = E3(nT) + E4(nT) + E5(nT)$ $\rightarrow$ $[1] = [1] + [0] + [0],$
1) $F2(nT) = F3(nT) = F4(nT) = F5(nT)$ $\rightarrow$ $[0] = [0] = [0] = [0],$
4) $E3(nT) = S1 \times F3(nT)$ $\rightarrow$ $[1] = [1] \times [0],$
2) $E4(nT) = I1 \times (F4(nT) - F4(n-1)T))$ $\rightarrow$ $[0] = [+] \times ([0] - [0]),$
4) $E5(nT) = F6(nT)$ $\rightarrow$ $[0] = [0],$
2) $F5(nT) = E6(nT)$ $\rightarrow$ $[0] = [0],$
5) $E6(nT) = E7(nT) + E8(nT) + E9(nT)$ $\rightarrow$ $[0] = [0] + [0] + [0],$
3) $F6(nT) = F7(nT) = F8(nT) = F9(nT)$ $\rightarrow$ $[0] = [0] = [0] = [0],$
4) $E7(nT) = R1 \times F7(nT)$ $\rightarrow$ $[0] = [+] \times [0],$
4) $E8(nT) = I2 \times (F8(nT) - F8((n-1)T))$ $\rightarrow$ $[0] = [+] \times ([0] - [0]),$
4) $E9(nT) = E10(nT)$ $\rightarrow$ $[0] = [0],$
2) $F9(nT) = F10(nT)$ $\rightarrow$ $[0] = [0],$
3) $E10(nT) = E11(nT) + E12(nT) + E24(nT)$ $\rightarrow$ $[0] = [0] + [0] + [0],$
1) $F10(nT) = F11(nT) = F12(nT) = F24(nT)$ $\rightarrow$ $[0] = [0] = [0] = [0],$
2) $E11(nT) = S2 \times F11(nT)$ $\rightarrow$ $[0] = [0] \times [0],$
1) $E12(nT) = R2 \times F12(nT)$ $\rightarrow$ $[0] = [+] \times [0],$
1) $E24(nT) = 0$ $\rightarrow$ $[0] = [0],$

In these equations, both the values of $E11(nT)$ and $E12(nT)$ are [0]. Also, it can be inferred that the values of $E22(nT)$ and $E23(nT)$ are [0]. That is, no variables violate the operational constraints. The shut-down sequence is thus obtained as: disconnecting S1 first and disconnecting S2 second.

7.6.3   The Emergency Measures Sequence

Suppose that the system has an abnormal behaviour where $E1(nT) = [0]$, $F12(nT) = [0]$, and $E30(nT) = [1]$ under the situation in which S1 and S2 are connected whilst S3 and S4 are disconnected. First, the values of the auxiliary measurement and controller output are inserted into the qualitative equations. Then, the value of the feedback measurement is set as [0] and inserted into the qualitative equations. Thus, the unknown variables can be obtained as follows:

1) $E1(nT) = E2(nT) = E13(nT)$ $\rightarrow$ $[0] = [0] = [0],$
8) $F1(nT) = F2(nT) + F13(nT)$ $\rightarrow$ $[0] = [0] + [0],$
5) $E2(nT) = E3(nT) + E4(nT) + E5(nT)$ $\rightarrow$ $[0] = [0] + [0] + [0],$
7) $F2(nT) = F3(nT) = F4(nT) = F5(nT)$ $\rightarrow$ $[0] = [0] = [0] = [0],$
8) $E3(nT) = S1 \times F3(nT)$ $\rightarrow$ $[0] = [0] \times [0],$

6) $E4(nT) = I1 \times (F4(nT) - F4(n-1)T))$ $\rightarrow$ $[0] = [+] \times ([0] - [0]),$

4) $E5(nT) = F6(nT)$ $\rightarrow$ $[0] = [0],$

6) $F5(nT) = E6(nT)$ $\rightarrow$ $[0] = [0],$

5) $E6(nT) = E7(nT) + E8(nT) + E9(nT)$ $\rightarrow$ $[0] = [0] + [0] + [0],$

3) $F6(nT) = F7(nT) = F8(nT) = F9(nT)$ $\rightarrow$ $[0] = [0] = [0] = [0],$

4) $E7(nT) = R1 \times F7(nT)$ $\rightarrow$ $[0] = [+] \times [0],$

4) $E8(nT) = I2 \times (F8(nT) - F8((n-1)T))$ $\rightarrow$ $[0] = [+] \times ([0] - [0]),$

3) $E9(nT) = E10(nT)$ $\rightarrow$ $[0] = [0],$

2) $F9(nT) = F10(nT)$ $\rightarrow$ $[0] = [0],$

2) $E10(nT) = E11(nT) + E12(nT) + E24(nT)$ $\rightarrow$ $[0] = [0] + [0] + [0],$

1) $F10(nT) = F11(nT) = F12(nT) = F24(nT)$ $\rightarrow$ $[0] = [0] = [0] = [0],$

2) $E11(nT) = S2 \times F11(nT)$ $\rightarrow$ $[0] = [0] \times [0],$

1) $E12(nT) = R2 \times F12(nT)$ $\rightarrow$ $[0] = [+] \times [0],$

1) $E24(nT) = 0$ $\rightarrow$ $[0] = [0],$

5) $E13(nT) = E14(nT) + E15(nT) + E16(nT)$ $\rightarrow$ $[0] = [0] + [0] + [0],$

1) $F13(nT) = F14(nT) = F15(nT) = F16(nT)$ $\rightarrow$ $[0] = [0] = [0] = [0],$

6) $E14(nT) = S3 \times F14(nT)$ $\rightarrow$ $[0] = [1] \times [0],$

2) $E15(nT) = I3 \times (F15(nT) - F15(n-1)T))$ $\rightarrow$ $[0] = [+] \times ([0] - [0]),$

4) $E16(nT) = F17(nT)$ $\rightarrow$ $[0] = [0],$

2) $F16(nT) = E17(nT)$ $\rightarrow$ $[0] = [0],$

5) $E17(nT) = E18(nT) + E19(nT) + E20(nT)$ $\rightarrow$ $[0] = [0] + [0] + [0],$

3) $F17(nT) = F18(nT) = F19(nT) = F20(nT)$ $\rightarrow$ $[0] = [0] = [0] = [0],$

4) $E18(nT) = R3 \times F18(nT)$ $\rightarrow$ $[0] = [+] \times [0],$

4) $E19(nT) = I4 \times (F19(nT) - F19((n-1)T))$ $\rightarrow$ $[0] = [+] \times ([0] - [0]),$

6) $E20(nT) = E21(nT)$ $\rightarrow$ $[0] = [0],$

2) $F20(nT) = F21(nT)$ $\rightarrow$ $[0] = [0],$

7) $E21(nT) = E22(nT) + E23(nT) + E25(nT)$ $\rightarrow$ $[0] = [0] + [0] + [0],$

1) $F21(nT) = F22(nT) = F23(nT) = F25(nT)$ $\rightarrow$ $[0] = [0] = [0] = [0],$

8) $E22(nT) = S4 \times F22(nT)$ $\rightarrow$ $[0] = [1] \times [0],$

2) $E23(nT) = R4 \times F23(nT)$ $\rightarrow$ $[0] = [+] \times [0],$

1) $E25(nT) = 0$ $\rightarrow$ $[0] = [0],$

7) $E26(nT) = E27(nT)$ $\rightarrow$ $[0] = [0],$

5) $F24(nT) + F25(nT) = F26(nT) + F27(nT)$ $\rightarrow$ $[0] + [0] = [0] + [0],$

6) $F26(nT) = C1 \times (E26(nT) - E26((n-1)T))$ $\rightarrow$ $[0] = [+] \times ([0] - [0]),$

6) $E27(nT) = E28(nT) + E29(nT)$ $\rightarrow$ $[0] = [0] + [0],$

4) $F27(nT) = F28(nT) = F29(nT)$ $\rightarrow$ $[0] = [0] = [0],$

5) $E28(nT) = R5 \times F28(nT)$ $\rightarrow$ $[0] = [+] \times [0],$

1) $E29(nT) = E30(nT) = E31(nT) = E32(nT)$ $\rightarrow$ $[0] = [0] = [0] = [0],$

3) $F29(nT) = F30(nT) + F31(nT) + F32(nT)$ $\rightarrow$ $[0] = [0] + [0] + [0],$

1) $F30(nT) = C2 \times (E30(nT) - E30((n-1)T))$ $\rightarrow$ $[0] = [+] \times ([0] - [0]),$

2) $E31(nT) = S5 \times F31(nT)$ $\rightarrow$ $[0] = [1] \times [0]$,

2) $E32(nT) = R6 \times F32(nT)$ $\rightarrow$ $[0] = [+] \times [0]$.

All the above equations are satisfied, which means that the abnormal behaviour is not caused by the switch states. Therefore, S1 and S2 can be kept connected. The next step is to find the power paths which are connected to the path Bonds-1, 2, 3, 4, 5, 6, 7, 8, 9, 10, 11, 12, 24, 26, 27, 28, 29, 30. Thus, the path Bonds-13, 14, 15, 16, 17, 18, 19, 20, 21, 22, 23, 25, and the path Bonds-31, 32, can be found. Then, the switches on the former path (S3 and S4) are assumed to be connected. It can be seen from the above inference that all the equations will be satisfied when S3 and S4 are connected. Therefore, connecting S3 and S4 is an acceptable emergency measure. Next, the switch in the path Bonds-31, 32 (S5) is assumed to be connected. Then, the unknown variables can be inferred. These inference results are very similar to the ones above (the only difference here is S5 = [0]), and all the equations are satisfied. Therefore, connecting S5 is also acceptable for emergency measures. Finally, through Steps 7) to 9) of the start-up sequence derivation procedure, the emergency measures can be obtained as: (1) connecting S4 first and then connecting S3, or (2) connecting S5. Both these two emergency measures are reasonable. Connecting S3 and S4 can prevent the faults occurring in the path Bonds-2, 3, 4, 5, 6, 7, 8, 9, 10, 11, 12, 24, while connecting S5 can decrease the liquid level in C2 effectively.

Again, it is supposed that the abnormal behaviour is observed as: $E1(nT) = [1]$, $E30(nT) = [0]$, and $F12(nT) = [0]$. Thus, the values of $F12(nT)$ and $E1(nT)$ are inserted into the qualitative equations, and then the value of $E30(nT)$ is set as $[1]$ and is inserted into the equations. Then, the unknown variables can be evaluated. From Eq. (7-22), the value of $F24$ can be obtained as $[0]$, and, from Eq. (7-39), the value of $F25$ is also $[0]$ (since S4 is disconnected). Thus, the inference from Eqs. (7-43) to (7-53) is shown as follows:

7) $E26(nT) = E27(nT)$ $\rightarrow$ $[-1] = [1]$,

5) $F24(nT) + F25(nT) = F26(nT) + F27(nT)$ $\rightarrow$ $[0] + [0] = [-1] + [1]$,

6) $F26(nT) = C1 \times (E26(nT) - E26((n-1)T))$ $\rightarrow$ $[-1] = [+] \times ([-1] - [0])$,

6) $E27(nT) = E28(nT) + E29(nT)$ $\rightarrow$ $[1] = [1] + [1]$,

4) $F27(nT) = F28(nT) = F29(nT)$ $\rightarrow$ $[1] = [1] = [1]$,

5) $E28(nT) = R5 \times F28(nT)$ $\rightarrow$ $[1] = [+] \times [1]$,

1) $E29(nT) = E30(nT) = E31(nT) = E32(nT)$ $\rightarrow$ $[1] = [1] = [1] = [1]$

3) $F29(nT) = F30(nT) + F31(nT) + F32(nT)$ $\rightarrow$ $[1] = [+] + [0] + [1]$,

1) $F30(nT) = C2 \times (E30(nT) - E30((n-1)T))$ $\rightarrow$ $[+] = [+] \times ([1] - [+])$,

2) $E31(nT) = S5 \times F31(nT)$ $\rightarrow$ $[1] = [1] \times [0]$,

2) $E32(nT) = R6 \times F32(nT)$ $\rightarrow$ $[1] = [+] \times [1]$.

It can be seen from the above inference that the equation $E26(nT) = E27(nT)$ is not satisfied. In this situation, if the fault candidates obtained through the fault diagnosis process contain the power driver and/or the controller, then this system should be shut down. Thus, the shut-down sequence can be produced as: disconnecting S1 first and then disconnecting S2. On the other hand, if the fault candidates do not include the power driver and the controller, then the switches in the power path which interlock the input and output variables (S1 and S2) should be disconnected. Next, the switches in the power paths which are connected to the path containing the fault candidates will be assumed to be connected. Thus, S3 and S4 are assumed to be connected, the switch states (S1 = [1], S2 = [1], S3 = [0], S4 = [0], and S5= [1]) are inserted into the qualitative equations., and the unknown variables can be inferred as follows:

1) $E1(nT)\ = E2(nT) = E13(nT)$ → $[1] = [1] = [1],$

19) $F1(nT)\ = F2(nT) + F13(nT)$ → $[1] = [0] + [1],$

5) $E2(nT)\ = E3(nT) + E4(nT) + E5(nT)$ → $[1] = [1] + [0] + [0],$

1) $F2(nT)\ = F3(nT) = F4(nT) = F5(nT)$ → $[0] = [0] = [0] =[0],$

6) $E3(nT)\ = S1 \times F3(nT)$ → $[1] = [1] \times [0],$

2) $E4(nT)\ = I1 \times (F4(nT) - F4(n-1)T))$ → $[0] = [+] \times ([0] - [0]),$

4) $E5(nT)\ = F6(nT)$ → $[0] = [0],$

2) $F5(nT)\ = E6(nT)$ → $[0] = [0],$

5) $E6(nT)\ = E7(nT) + E8(nT) + E9(nT)$ → $[0] = [0] + [0] + [0],$

3) $F6(nT)\ = F7(nT) = F8(nT) = F9(nT)$ → $[0] = [0] = [0] = [0],$

4) $E7(nT)\ = R1 \times F7(nT)$ → $[0] = [+] \times [0],$

4) $E8(nT)\ = I2 \times (F8(nT) - F8((n-1)T))$ → $[0] = [+] \times ([0] - [0]),$

6) $E9(nT)\ = E10(nT)$ → $[0] = [0],$

2) $F9(nT)\ = F10(nT)$ → $[0] = [0],$

7) $E10(nT) = E11(nT) + E12(nT) + E24(nT)$ → $[0] = [0] + [0] + [0],$

1) $F10(nT) = F11(nT) = F12(nT) = F24(nT)$ → $[0] = [0] = [0] = [0],$

8) $E11(nT) = S2 \times F11(nT)$ → $[0] = [1] \times [0],$

2) $E12(nT) = R2 \times F12(nT)$ → $[0] = [+] \times [0],$

1) $E24(nT) = 0$ → $[0] = [0],$

20) $E13(nT) = E14(nT) + E15(nT) + E16(nT)$ → $[1] = [0] + [+] + [1],$

18) $F13(nT) = F14(nT) = F15(nT) = F16(nT)$ → $[1] = [1] = [1] = [1],$

19) $E14(nT) = S3 \times F14(nT)$ → $[0] = [0] \times [1],$

19) $E15(nT) = I3 \times (F15(nT) - F15(n-1)T))$ → $[+] = [+] \times ([1] - [+]),$

15) $E16(nT) = F17(nT)$ → $[1] = [1],$

17) $F16(nT) = E17(nT)$ → $[1] = [1],$

16) $E17(nT) = E18(nT) + E19(nT) + E20(nT)$ → $[1] = [+] + [1] + [1],$

14) $F17(nT) = F18(nT) = F19(nT) = F20(nT)$ → $[1] = [1] = [1] = [1],$

15) $E18(nT) = R3 \times F18(nT)$ $\rightarrow$ $[1] = [+] \times [1],$
15) $E19(nT) = I4 \times (F19(nT) - F19((n-1)T))$ $\rightarrow$ $[+] = [+] \times ([1] - [+]),$
14) $E20(nT) = E21(nT)$ $\rightarrow$ $[1] = [1],$
13) $F20(nT) = F21(nT)$ $\rightarrow$ $[1] = [1],$
13) $E21(nT) = E22(nT) + E23(nT) + E25(nT)$ $\rightarrow$ $[1] = [0] + [1] + [0],$
11) $F21(nT) = F22(nT) = F23(nT) = F25(nT)$ $\rightarrow$ $[1] = [1] = [1] = [1],$
12) $E22(nT) = S4 \times F22(nT)$ $\rightarrow$ $[0] = [0] \times [1],$
12) $E23(nT) = R4 \times F23(nT)$ $\rightarrow$ $[1] = [+] \times [1],$
1) $E25(nT) = 0$ $\rightarrow$ $[0] = [0],$
7) $E26(nT) = E27(nT)$ $\rightarrow$ $[1] = [1],$
9) $F24(nT) + F25(nT) = F26(nT) + F27(nT)$ $\rightarrow$ $[0] + [1] = [+] + [1],$
8) $F26(nT) = C1 \times (E26(nT) - E26((n-1)T))$ $\rightarrow$ $[+] = [+] \times ([1] - [+]),$
6) $E27(nT) = E28(nT) + E29(nT)$ $\rightarrow$ $[1] = [1] + [1],$
4) $F27(nT) = F28(nT) = F29(nT)$ $\rightarrow$ $[1] = [1] = [1],$
5) $E28(nT) = R5 \times F28(nT)$ $\rightarrow$ $[1] = [+] \times [1],$
1) $E29(nT) = E30(nT) = E31(nT) = E32(nT)$ $\rightarrow$ $[1] = [1] = [1] = [1],$
3) $F29(nT) = F30(nT) + F31(nT) + F32(nT)$ $\rightarrow$ $[1] = [+] + [0] + [1],$
1) $F30(nT) = C2 \times (E30(nT) - E30((n-1)T))$ $\rightarrow$ $[+] = [+] \times ([1] - [+]),$
2) $E31(nT) = S5 \times F31(nT)$ $\rightarrow$ $[1] = [1] \times [0],$
2) $E32(nT) = R6 \times F32(nT)$ $\rightarrow$ $[1] = [+] \times [1].$

It can be seen from this inference that all the equations are satisfied. Therefore, connecting S3 and S4 and disconnecting S1 and S2 are acceptable as emergency measures. Then, S5 is assumed to be connected. In this situation, the value of the effort of S5 ($E31$) is [0], whereas the value of output variable $E30$ is [1]. Thus, it can be inferred that Eq. (7-49) will not be satisfied. This is to say that connecting S5 is unacceptable. Finally, through Steps 4) and 5) of the shut-down sequence derivation procedure and Steps 7) to 9) of the start-up sequence derivation procedure, the emergency measures sequence is obtained as: connecting S4 and disconnecting S1 first; and then connecting S3 and disconnecting S2. In this example, the assumed abnormal behaviour ($E1(nT) = [1]$, $F12(nT) = [0]$, and $E30(nT) = [0]$) can be caused by a leakage in the connected power path or a blockage of the motor and pump. Thus, disconnecting the faulty power path (by disconnecting S1 and S2) and connecting the spare power path (by connecting S3 and S4) may correct the abnormal behaviour. The emergency measures sequence generated above is quite reasonable.

7.7 DISCUSSION

This chapter has proposed an intelligent supervisory control system, where the hybrid qualitative and quantitative control method, the auto-tuning scheme, and the qualitative fault diagnosis method developed in the previous chapters have been integrated to cope with different classes of system behaviours. A management mechanism has been developed to schedule the appropriate control regimes for various system behaviours through performance monitoring. The experiments show that this supervisory control system can supervise a MIMO coupled tanks apparatus successfully. Further, the qualitative representation has been extended to describe the behaviour of discontinuous switches. Based on this extended representation, the sequences of start-up, shut-down, and emergency measures for a system can be generated through cause-effect inference to support decision making.

In summary, this intelligent supervisory control system consists of generic control tasks to achieve the control goal given by the user. At the beginning, when the knowledge about a process (such as the process structure, control goal, operational constraints, performance criteria, and the locations of input and output variables) has been given, a number of start-up sequences can be derived to suggest how to start the process. Then, after the process has started, the feedback controller begins to regulate the process, and the auto-tuning mechanism will then find the suitable scaling factors to keep the process behaviour within the performance criteria. If an abnormal behaviour occurs during the process, the auto-tuning mechanism will be activated first to try to bring the behaviour back into the acceptable region. If this does not work, then the fault diagnosis mechanism will be activated to localise possible process faults, and then a number of emergency measures sequences will be suggested by the planner according to the diagnosis result to help the user manage the abnormal behaviour. Finally, if the process needs to be shut down, the planner can suggest a shut-down sequence to help stop the process safely.

A real engineering system usually contains a large amount of logical switches and may be operated in a number of different modes. Accordingly, a lot of different system models and control algorithms are required for supervising the system in various modes. These models and control algorithms can be generated off-line at the design stage and stored in a knowledge base. Thus, the management mechanism can quickly adopt an appropriate model and control algorithm for an individual control mode and thus keep the on-line operation efficient.

This supervisory control method incorporates different control regimes into a uniform framework. The deep models used in this supervisory system can be built easily through the generalised qualitative bond graph modelling scheme. All the structural information required for control algorithm derivation, fault diagnosis, and operation sequence derivation can be acquired from the models directly. A further possibility, which has been discussed in Chapter 3, is to model a system and derive control algorithms automatically. Thus, even higher automation can be achieved.

CHAPTER 8

CONCLUSIONS AND
SUGGESTIONS FOR FUTURE WORK

8.1 SUMMARY OF THE APPROACH

In Chapter 1, the motivation for this research and the background to the use of qualitative bond graph reasoning were presented. The purpose of this approach was to develop an easily usable real-time supervisory control system to provide a robust semi-autonomous system using qualitative design. Successive chapters have proposed a number of methodologies for qualitative representation of dynamic systems, qualitative bond graph modelling, hybrid qualitative and quantitative control, auto-tuning, fault diagnosis, operation sequence derivation, and system supervision. Implementations of these methods have been illustrated by experiments on the SISO and MIMO coupled tanks liquid level control rigs, showing that qualitative techniques can be applied successfully to real-time control tasks.

In this approach, bond graphs provide the capability to represent cross-domain engineering systems with a unified language, while qualitative reasoning allows the inference to be performed when system information is incomplete. The combination of qualitative reasoning with bond graphs makes it possible to build an intelligent control system which can acquire knowledge by itself through a modelling procedure rather than collecting information from different human experts, and can regulate a process based on the understanding of the process structure without evaluating its parameters. The control action of such a system is very close in spirit to the behaviour of human operators. Human intervention can be further reduced in both controller design and plant operation by applying this intelligent control method.

191

The approaches developed in this book have been demonstrated with a number of examples drawn from different fields (electrical, electro-mechanical, and electro-fluid). These qualitative methods have a great deal of generality due to the use of bond graphs, in which cross-field dynamic systems can be modelled with a limited number of physical primitives. Therefore, systems which can be modelled by bond graphs can be regulated by the qualitative-bond-graph-model-based supervisory control system. However, more examples should be studied to examine the generality of the intelligent control method.

Qualitative reasoning is usually classified as belonging to the field of deep-level knowledge reasoning. Morgan [1988] has stated that "depth" in the reasoning comes from the use of a supporting model. Most approaches in this field have defined that a deep model is one which describes the system structure, components' behaviour, and the functional relations between components. Against this, a shallow model is the model built on empirical associations between input data and output solutions. Another viewpoint about the deep model which has emerged from the work proposed in this book is that "a deep model is a model built independently without aiming at specific applications, and providing different aspects of knowledge for different applications". In conventional mathematical and rule-based approaches, modelling is task-oriented. Scientists and engineers abstract necessary knowledge from a system to build models for different purposes using different knowledge representations (e.g. transfer functions, ordinary differential equations, heuristic rules, etc.). Therefore, a model developed for one application cannot be used by other applications. For example, a model built for controller design is very difficult to be used for fault diagnosis. Such models can be seen as "shallow". In this book, qualitative bond graph models were constructed to represent system structures without abstraction for individual purposes. Therefore, these models contain a large amount of knowledge needed for various applications and can thus be called "deep models". Different control tasks (controller design, fault diagnosis, and operation sequence derivation) in the supervisory control system can be performed based on a single deep model.

In contrast to most qualitative reasoning approaches, applications developed in this book did not employ the notion of causality to resolve problems. Causality is usually applied to two aspects in qualitative reasoning: modelling and simulation. At the modelling stage, the causality is used to combine the algebraic relations between variables of different components. At the simulation stage, the causality is used to reason about which variables will be affected by a input signal and then evaluate these variables. In this book, variables were related in a model according to the interconnections between components without taking account of causalities. On the other hand, this book did not use simulations for cause-effect inferences to predict unknown variables. In the analysis tasks (fault diagnosis and operation sequence derivation), unknown system variables were evaluated through resolving qualitative equations using the known system input and output and the hypothesis of system state. Causality is not needed for equation-solving. The work proposed in this book shows that "causality is not necessary for some qualitative applications". A real-time qualitative system can be implemented more easily without spending time in causality-analysis.

8.2   SUGGESTIONS FOR FUTURE WORK

Much of the capability of the supervisory control system comes from the use of qualitative bond graph models. Models which provide more features of controlled systems will further improve the performance of the supervisory control system. Enhancing the capability of the models is the most direct way to enhance the capability of the supervisory control system. Therefore, future work on this approach should be concentrated on resolving the problems of qualitative bond graph models.

Determining the Detail-Level of Qualitative Models:

The most difficult problem in this approach is developing an automatic method to determine the appropriate detail-level of a qualitative bond graph model. The detail-level decides the accuracy of a qualitative model. In bond graphs, basic elements (resistance, capacitance, transformatance, etc.) used to represent dynamic systems are defined as ideal. A realistic physical system is usually composed of a number of ideal elements. In a numerical model, the importance of an element to the features of the system is represented by the parameter of the element. A significant element has a big parameter, and *vice versa*. However, in a qualitative bond graph model, there are no numerical parameters to indicate the importance of elements. Therefore, the effect of an insignificant element could be exaggerated in the qualitative analysis so as to result in an erroneous inference. At this point, the most accurate qualitative model could be the one in which insignificant elements are ignored, rather than the most detailed one which contains all the system elements. How to determine the significance of an element is a critical problem.

A direct solution to this problem is to use system identification techniques to evaluate the parameters of system elements so that their importance to the features of a system can be identified. However, this method could be expensive and time-consuming. Another method developed in qualitative reasoning is "time scale abstraction" [Rickel and Porter, 1992]. This abstraction allows the detailed dynamics of a process much faster than the time scale of interest to be abstracted out. However, this method requires the user to specify all the important time scales (such as seconds, minutes, or days) at which a system works, and classify the system behaviour according to these time scales. When the system is complex, it is difficult for the user to meet these requirements. A general method for the detail-level determination is one of the most important tasks in improving the ability of qualitative bond graph modelling.

Enhanced Qualitative Representation:

The qualitative representation of dynamic systems proposed in this book has described successfully the behaviour of continuous components, discontinuous switches, and feedback components. However, two common features of industrial processes, non-linearity and time-delay, have not been covered by this qualitative representation.

In qualitative reasoning, a measurement space is represented by a simple set of qualitative values (such as {1, +, 0, -, -1} used in this book). These qualitative descriptors can be used to represent linear and monotonic behaviours but they are not sufficient to describe the non-linear behaviour. The value of a quantity is either positive or negative (or exactly zero). This value cannot describe the non-linear changes of a quantity. A possible solution to this problem is the piecewise linear approximation. A non-linear system can be represented by a number of linear qualitative models. Each linear model describes the local properties of the system. In the supervisory control system of this book, the management mechanism could be used to choose suitable models to cope with various local behaviours of a non-linear system.

The difficulty of representing time-delays in this approach is caused by the formulation of qualitative equations. A realistic component in bond graphs is composed of several basic ideal elements, and qualitative equations are used to represent the functions of these ideal elements. The overall function of a component is described by a set of qualitative equations. On the other hand, the time-delay of a component can be measured by observing the time at which the output begins to respond to a given input. However, it is difficult to determine which basic element causes the time delay. In other words, it is hard to find the "correct" equation from a set of qualitative equations to describe the time-delay. This problem is not very serious for the feedback control task, because the knowledge required by a feedback controller is only the relations between the system input and output. The total time delay of a system can be represented as a time interval between the system input and output. However, this problem can be serious for fault diagnosis. The fault diagnosis method proposed in this book needs to analysis the interrelations of all system variables. An ill-defined time-delay representation may describe wrong relations between the variables and cause a wrong diagnosis result. How to represent time-delays correctly is an important issue in this approach.

Representation of Uncertainty:

Ambiguities generated from qualitative operations limit the inference capability of the intelligent control system. For example, the fault diagnosis mechanism could suggest too many fault candidates because of the uncertain situations of internal variables. One solution to this problem is introducing the concepts of stochastic and fuzzy set theories to represent uncertain values. Shen and Leitch [1993] have integrated fuzzy logic and qualitative reasoning techniques and successfully avoided the difficulties of representing uncertainty and time (as discussed in Chapter 2). However, this method requires further human intervention to assign membership functions. A method which can describe uncertainty objectively will improve the qualitative inference mechanisms.

The overall conclusion becomes that the "model" is the key to the success of an intelligent supervisory control system. The amount of knowledge in a model determines the problem-solving power. In a good model, knowledge is organised so that it can be expressed systematically; knowledge is dynamic so that it can be applied appropriately to a specific problem; and knowledge is flexible so that it can be applied freely to a class of problems. A qualitative bond graph model largely satisfies the conditions of a good model, so that an intelligent supervisory control system can be achieved. The determination of limitations on qualitative bond graph models and their solution will contribute towards higher automation for industrial control.

REFERENCES

Abbod M. F., 1992, "Supervisory Intelligent Control for Industrial and Medical Systems", Ph.D. Thesis, University of Sheffield, England.

Årzén K.-E., 1988, "An Architecture for Expert System Based Feedback Control", *Automatica*, Vol. 25, No. 6, pp. 813 - 827.

Åström K. J., J. J. Anton, and K.-E. Årzén, 1986, "Expert Control", *Automatica*, Vol. 22, No. 3, pp. 277 - 286.

Åström K. J. and B. Wittenmark, 1989, *Adaptive Control*, Addison-Wesley, Wokingham, UK.

Barreto J. M., 1988. "The Role of Bond Graphs in Qualitative Modelling", *Proceedings of the 12th IMACS World Congress on Scientific Computation*, Vol. 1, pp. 84 - 87.

Biswas G., X. Yu, and K. Debelak, 1992, "A Formal Modelling Scheme for Continuous-Valued Systems: Focus on Diagnosis", *The Sixth International Workshop on Qualitative Reasoning about Physical Systems*, Heriot-Watt Univ., Edinburgh, pp. 302 - 321.

Bobrow D. G. (ed.), 1984, Qualitative Reasoning about Physical Systems, *Special Issue of Artificial Intelligence*, Vol. 24, North-Holland, Amsterdam.

Broenink J. F. and G. D. Nijen Twilhaar, 1985, "CAMAS, A Computer Aided Modelling, Analysis and Simulation Environment", *Proceedings of the 4th International Conference on Engineering Software IV*, pp. 15 - 16.

Cellier F. E. and J. J. Granda (ed.), 1995, *Proceedings of the International Conference on Bond Graph Modelling and Simulation* (ICBGM'95), The Society for Computer Simulation, Las vegas, USA.

Cheung J. T.-Y. and G. Stephanopoulos, 1990a, "Representation of Process Trends — Part I. A Formal Representation Framework", *Computers Chemical Engineering*, Vol. 14, No. 4/5, pp. 495 - 510.

Cheung J. T.-Y. and G. Stephanopoulos, 1990b, "Representation of Process Trends — Part II. The Problem of Scale and Qualitative Scaling", *Computers Chemical Engineering*, Vol. 14, No. 4/5, pp. 511 - 539.

Clocksin W. F. and A. J. Morgan, 1986, "Qualitative Control", *Proceedings of the 7th European Conference on Artificial Intelligence*, Brighton, UK, Vol. 1, pp. 350 - 356.

Davis R. and D. B. Lenat, 1982, *Knowledge-Based Systems in Artificial Intelligence*, McGraw-Hill, New York.

Davis R., 1984, "Diagnostic Reasoning Based on Structure and Behaviour", *Artificial Intelligence*, Vol. 24, pp. 347 - 410.

de Kleer J., 1977, "Multiple Representations of Knowledge in a Mechanics Problem-Solver", *Proceedings of IJCAI-77*, pp. 299 - 304.

de Kleer J. and D. G. Bobrow, 1984, "Qualitative Reasoning with Higher-order Derivatives", *Proceedings of AAAI-84*, pp. 86 - 91.

de Kleer J. and J. S. Brown, 1984, "A Qualitative Physics Based on Confluences", *Artificial Intelligence*, Vol. 24, pp. 7 - 84.

de Kleer J. and J. S. Brown, 1986, "Theories of Causal Ordering", *Artificial Intelligence*, Vol. 29, pp 33 - 61.

de Kleer J. and B. C. Williams, 1987, "Diagnosing Multiple Faults", *Artificial Intelligence*, Vol. 32, pp. 97 - 130.

de Kleer J., 1993, "A View on Qualitative Physics", *Artificial Intelligence*, Vol. 59, pp. 105 - 114.

de Silva C. W. and A. G. Macfarlane, 1989, "Knowledge-based Control Approach for Robotic Manipulator", *International Journal of Control*, Vol. 50, No. 1, pp. 249 - 273.

Far B. H., 1989, "Qualitative Control: Qualitative Design and Verification of Industrial Controllers", *International Workshop on Industrial Applications of Machine Intelligence and Vision MIV-89*, Tokyo, pp. 239 - 244.

Falkenhainer B. and K. D. Forbus, 1991, "Compositional Modelling: Finding the Right Model for the Job", *Artificial Intelligence*, Vol. 51, pp. 95 - 143.

Falkenhainer B., 1992, "Modelling without Amnesia: Making Experience-sanctioned Approximations", *The Sixth International Workshop on Qualitative Reasoning about Physical Systems*, Heriot-Watt Univ., Edinburgh, Scotland, pp. 44 - 55.

Filippo J. M.-D., C. Brid, and H. M. Paynter, 1991, "A Survey of Bond Graphs: Theory, Applications and Programs", *Journal of the Franklin Institute*, Vol. 328, No. 5/6, pp. 565 - 606.

Forbus K. D., 1984, "Qualitative Process Theory", *Artificial Intelligence*, Vol. 24, pp. 85 - 168.

Forbus K. D., P. Nielsen, and B. Faltings, 1987, "Qualitative Kinematics: A Framework", *Proceedings of IJCAI-87*, pp. 430 - 436.

Forbus K. D. and B. Falkenhainer, 1992, "Self-Explanatory Simulations: Scaling Up to Large Model", *The Sixth International Workshop on Qualitative Reasoning about Physical Systems*, Heriot-watt Univ., Edinburgh, Scotland, pp. 22 -35.

Francis J. C. and R. R. Leitch, 1984, "ARTIFACT: A Real-time Shell for Intelligent Feedback Control", *Research and Development in Expert System*, Edited by M. Bramer, Cambridge Univ. Press, pp. 151 - 162.

Frank P. A., 1990, "Fault Diagnosis in Dynamic System Using Analytical and Knowledge-based Redundancy — A Survey and Some New Results", *Automatica*, Vol. 26, No. 3, pp. 459 - 474.

Franklin G. F., J. D. Powell, and A. Emami-Naeini, 1986, *Feedback Control of Dynamic Systems*, Addison-Wesley, Taipei.

Fu K. S., 1971, "Learning Control Systems and Intelligent Control Systems: An Intersection of Artificial Intelligence and Automatic Control", *IEEE Transactions on Automatic Control*, Vol. AC-16, pp. 70 - 72.

Genesereth M. R., 1984, "The Use of Design Descriptions in Automated Diagnosis", *Artificial Intelligence*, Vol. 24, pp. 411 - 436.

Granda J. J., 1985, "Computer Generation of Physical System Differential Equations Using Bond Graphs", *Journal of the Franklin Institute*, Vol. 319, No. 1/2, pp. 243 - 255.

Gupta M. M., A. Kandel, W. Bandler, and J. B. Kiszka, (ed.), 1985, *Approximate Reasoning in Expert Systems*, North-Holland, Amsterdam.

Handelman D. A., S. H. Lane, and J. J. Gelfand, 1988, "Integration of Knowledge-based System and Neural Network Techniques for Robotic Control", *Proceedings of IFAC Workshop on Artificial Intelligence in Real-time Control*, Swansea.

Haton J. P., 1983, "Knowledge-based and Expert Systems in Industrial Applications", *IFAC Workshop on Artificial Intelligence*, Lningrad, pp. 83 - 89.

Hayes P. J., 1979, "The Naive Physics Manifesto", *Expert Systems in the Microelectronics Age*, D. Michie ed., Edinburgh University Press.

Hayes P. J., 1985, "The Second Naive Physics Manifesto", *Formal Theories of the Commonsense World*, J. R. Hobbs and R. C. Moore (eds.), Ablex Publishing Corporation, pp. 1 - 36.

Hood S. J., E. R. Palmer, and P. M. Dantzig, 1989, "A Fast, Complete Method for Automatically Assigning Causality to Bond Graphs", *Journal of the Franklin Institute*, Vol. 326, No. 1, pp. 83 - 92.

Infelise N., 1991, "A Clear Vision of Fuzzy Logic", *Control Engineering*, Vol. 38, No. 9, pp. 28 - 30.

Iwasaki Y. and H. A. Simon, 1986, "Causality in Device Behaviour", *Artificial Intelligence*, Vol. 29, pp. 3 - 32.

Karnopp D. C., 1975, "Some Bond Graph Identities Involving Junction Structure", *Transactions of the ASME Journal of Dynamic Systems, Measurement, and Control*, Vol. 97, No. 4, pp. 439 - 440.

Kaufmann A., 1975, *Introduction to Theory of Fuzzy Subsets*, Academic, New York.

Kaufmann A. and M. M. Gupta, 1985, *Introduction to Fuzzy Arithmetic*, Van Nostrand, New York.

Klir G. J. and T. A. Folger, 1988, *Fuzzy Sets, Uncertainty, and Information*, Prentice-Hall, London.

Kruse R., 1984, "Statistical Estimation with Linguistic Data", *Information Science*, Vol. 33, pp. 197 - 207.

Kraus T. W. and T. J. Myron, 1984, " Self-Tuning PID Controller Uses Pattern Recognition Approach", *Control Engineering*, pp. 106 - 111.

Kuipers B. J., 1984, "Commonsense Reasoning about Causality: Deriving Behaviour from Structure", *Artificial Intelligence*, Vol. 24, pp. 169 - 203.

Kuipers B. J., 1986, "Qualitative Simulation", *Artificial Intelligence*, Vol. 29, pp. 289 - 338.

Kuipers B. J., 1993, "Reasoning with Qualitative Models", *Artificial Intelligence*, Vol. 59, pp. 125 - 132.

Kwakernaak H., 1978, "Fuzzy Random Variables. Part I: Definitions and Theorems", *Information Science*, Vol. 15, pp. 1 - 15, 1978.

Kwakernaak H., 1979, "Fuzzy Random Variables. Part II: Algorithms and Examples for the Discrete Case", *Information Science*, Vol. 17, pp. 253 - 278.

Lackinger F. and W. Nejdl, 1991, "Integrating Model-Based Monitoring and Diagnosis of Complex Dynamic Systems", *The Fifth International Workshop on Qualitative Reasoning about Physical Systems*, Austin, Texas, pp. 151 -170.

Leitch R. R. and C. Quek, 1992, "Architecture for Integrated Process Supervision", *IEE Proceedings-D*, Vol. 139, No. 3, pp. 317 - 327.

Leyval L., S. Gentil, and S. Feray-Beaumont, 1994, "Model-based Causal Reasoning for Process Supervision", *Automatica*, Vol. 30, No. 8, pp. 1295 - 1306.

Linkens D. A., S. Xia, and S. Bennett, 1991, "QREMS: Qualitative Reasoning Environment for Modelling and Simulation", *IEE Colloquium on Model Building Aids for Dynamic System Simulation*, Coventry, UK, pp. 1 - 5.

Linkens D. A., S. Xia, and S. Bennett, 1993, "A Computer-Aided Qualitative Modelling and Analysis Environment Using Unified Principles (QREMS)", *Proceedings of International Conference on Bond Graph Modelling ICBGM '93*, Simulation Series Vol. 25, No. 2, Lajolla, CA, pp. 53 - 58.

Liu K., J. Gertler, H. Chizek, and P. Katona, 1987, "A Supervisory (Expert) Adaptive Control Scheme", *Proceedings of 10th World Congress of IFAC*, Munich, pp. 375 - 380.

Lorenz I. F., 1993, "Discontinuities in Bond Graphs: What is Required?", *Proceedings of International Conference on Bond Graph Modelling ICBGM '93*, Simulation Series Vol. 25, No. 2, Lajolla, CA, pp. 137 - 142.

Luger G. F. and W. A. Stubblefield, 1989, *Artificial Intelligence and the Design of Expert Systems*, Benjamin/Cummings, California.

Martens H. R., 1973, "Simulation of Nonlinear Multiport Systems Using Bond Graphs", *Transactions of the ASME Journal of Dynamic Systems, Measurement, and Control*, Vol. 95, pp. 49 - 53.

Mavrovouniotis M. L. and G. Stephanopoulos, 1988, "Formal Order-of-Magnitude reasoning in Process Engineering", *Computers Chemical Engineering*, Vol. 12, pp. 867 - 880.

Morgan A. J., 1988, "The Qualitative Behaviour of Dynamic Physical Systems", Ph.D. Thesis, Wolfson College, Univ. of Cambridge, England.

Murthy S. V. and A. Kandel, 1990, "Fuzzy Sets and Typicality Theory", *Information Science*, Vol. 51, pp. 61 - 93.

Newell A. and H. Simon, 1976, "Computer Science as Empirical Inquiry: Symbols and Search", *Communications of the ACM*, Vol. 19, No. 3, pp. 113 - 126.

Ng H. T., 1990, "Model-based, Multiple-fault diagnosis of Dynamic, Continuous Physical Devices", *IEEE Expert*, pp. 38 - 43.

Nilsson N. J., 1969, "A Mobile Automaton: an Application of Artificial Intelligence Techniques", *Proceedings of 1969 International Joint Conference on Artificial Intelligence*.

Paynter H. M. 1959, "Hydraulics by analog - An Electronic Model of a Pumping Plant", *Journal of Boston Society of Civil Engineering*, July, pp. 197 - 219.

Perng M. H. and H. H. Chang, 1993, "Intelligent Supervision of Servo Control", *IEE Proceedings-D*, Vol. 140, No. 6, pp. 405 - 412.

Procyk T. J. and E. H. Mamdani, 1979, "A Linguistic Self-Organising Process Controller", *Automatica*, Vol. 15, pp. 15 - 30.

Ravindranathan M and R. Leitch, "Preliminary Classification of Architectures for Intelligent Systems", The Second International Conference on Intelligent Systems (SPICIS '94), 14 - 17 November, Singapore, pp. 1320 - 132.

Raiman O., 1986, "Order of Magnitude Reasoning", *Proceedings of the Fifth National Conference on Artificial Intelligence (AAAI-86)*, pp. 100 - 104.

Rasmussen J., 1986, *Information Processing and Human Machine Interaction, An Approach to Cognitive Engineering*, Series in System Science and Engineering, North-Holland.

Reiter R., 1987, "A Theory of Diagnosis from First Principles", *Artificial Intelligence*, Vol. 32, pp. 57 - 95.

Rickel J. and B. Porter, 1992, "Automated Modelling for Answering Prediction Questions: Exploiting Interaction Paths", *The Sixth International Workshop on Qualitative Reasoning abut Physical Systems*, Heriot-Watt Univ., Edinburgh, pp. 82 - 95.

Rosenberg R. C., 1965, "Computer Aided Teaching of Dynamic System Behaviour", Sc. D. Thesis, Department of Mechanical Engineering, M. I. T.

Rosenberg R. C., 1971, "State-Space Formulation for Bond Graph Models of Multiport Systems", *Transactions of the ASME Journal of Dynamic Systems, Measurement, and Control*, Vol. 93, No. 1, pp. 35 - 40.

Rosenberg R. C. and D. C. Karnopp, 1972, "A Definition of the Bond Graph Language", *Transactions of the ASME Journal of Dynamic Systems, Measurement, and Control*, Vol. 94, No. 3, pp. 179 - 182.

Rosenberg R. C., 1973, "Modelling and Simulation of Large-Scale, Linear Multiport Systems", *Automatica*, Vol. 9, pp. 87 - 95.

Rosenberg R. C. and D. C. Karnopp, 1983, *Introduction to Physical System Dynamics*, McGraw-Hill, New York.

Saridis G. A. and . H. E. Stephanou, 1977, "A Hierarchical Approach to the Control of Prosthetic Arm", *IEEE Transactions on System, Man, and Cybernetics*, Vol. SMC-7, No. 6, PP. 407 - 420.

Saridis G. A., 1983, "Intelligent Robotic Control", *IEEE Transactions on Automatic Control*, Vol. AC-28, No. 5, pp. 547 - 556.

Saridis G. A., 1989, "Analytic Formulation of the Principle of Increasing Precision with Decreasing Intelligence for Intelligent Machines", *Automatica*, No. 25, pp. 461 - 467.

Shen Q. and R. R. Leitch, 1990, "A Semi-Quantitative Extension to Qualitative Simulation*", Tenth International Workshop on Expert System and Their Applications General Conference, Second Generation of Expert Systems*, Arignon, France, pp. 223 - 234.

Shen Q. and R. R. Leitch, 1992, "Qualitative Model Based Diagnosis of Continuous Dynamic Systems", *Proceedings of the First International Conference on Intelligent Systems Engineering*, Heriot-Watt University, Edinburgh, UK, pp. 147 - 152.

Shen Q. and R. R. Leitch, 1993, "Fuzzy Qualitative Simulation", *IEEE Transactions on System, Man, and Cybernetics*, Vol. 23, No. 4, pp. 1038 - 1061.

Skorstad G. and K. D. Forbus, 1989, "Qualitative and Quantitative Reasoning about Thermodynamics", *Proceedings Eleventh Annual Conference of the Cognitive Science Society*. Ann Arbor, MI, pp. 412 - 420.

Struss P., 1988, "Mathematical Aspects of Qualitative Reasoning", *International Journal for Artificial Intelligence in Engineering*, Vol. 3, No. 3, pp. 156 - 169.

Struss P., 1990, "Problems of Interval-based Qualitative Reasoning", D. S. Weld and J. de Kleer, eds., *Readings in Qualitative Reasoning about Physical Systems*, Morgan Kaufmann, San Mateo, CA, pp. 288 - 305.

Top J. L., J. M. Akkermans and P. C. Breedveld, 1991, "Qualitative Reasoning about Physical System: An Artificial Intelligence Perspective", *J. Franklin Institute*, Vol. 328, No. 5/6, pp. 1047-1065.

Voss H., 1988, "Architectural Issues for Expert Systems in Real-Time Control", *Proceedings of IFAC Workshop on Artificial Intelligence in Real-Time Control*, Swansea, pp. 1 - 6.

Wang P. P. (ed.), 1983, *Advances in Fuzzy Sets, Possibility Theory, and Applications*, Plenum Press, New York.

Weld D. S. and J. de Kleer (ed.), 1990, *Readings in Qualitative Reasoning about Physical Systems*, Morgan Kaufmann, San Mateo.

Werthner H., 1994, *Qualitative Reasoning—Modelling and the Generation of Behaviour*, Springer-Verlag, Wien.

Williams B. C., 1984, "Qualitative Analysis of MOS Circuits", *Artificial Intelligence*, Vol. 24, pp. 281 - 346.

Williams B. C., 1986, "Doing Time: Putting Qualitative Reasoning on Firmer Ground", *Proceedings of the Fifth National Conference on Artificial Intelligence AAAI-86*, pp. 105 - 112.

Williams B. C., 1990, "Temporal Qualitative Analysis: Explaining How Physical Systems Work", *Readings in Qualitative Reasoning about Physical Systems*, D. S. Weld and J. de Kleer (eds.), Morgan Kaufmann, San Mateo, pp. 133 - 177.

Williams B. C., 1991, "A Theory of Interactions: Unifying Qualitative and Quantitative Algebraic Reasoning", *Artificial Intelligence*, Vol. 51, pp. 39 - 94.

Williams B. C. and J. de Kleer, 1991, "Qualitative Reasoning About Physical Systems: A Return to Roots", *Artif. Intell.*, Vol. 51, pp. 1 - 9.

Xia S., D. A. Linkens and S. Bennett, 1991, "Qualitative Reasoning and Applications to Dynamic Systems: A Bond Graph Approach", *Decision Support Systems and Qualitative Reasoning. Proceedings of IMACS International Workshop*, Toulouse, France, pp. 175 - 180.

Young P. C., 1984, *Recursive Estimation and Time-Series Analysis*, Springer Verlag, Berlin.

Zadeh L. A., 1965, "Fuzzy Set", *Information Control*, Vol. 8, pp. 338 - 353, 1965.

Zadeh L. A., 1975, "The Concept of a Linguistic Variable and its Application to Approximate Reasoning", *Information Sciences*, Vol. Part I, 8 pp. 199 - 249; Part II, 8, pp. 301 - 357; Part III, 9, pp. 43 - 80.

Ziegler J. G. and N. B. Nichols, 1942, "Optimum Settings for Automatic Controllers", *Transactions of ASME 64*, pp. 759 - 768.

INDEX